TransEvolution

The Coming Age
of Human Deconstruction

Daniel Estulin

TransEvolution: The Coming Age of Human Deconstruction
Copyright © 2014 Daniel Estulin. All Rights Reserved.
Presentation Copyright © 2014 Trine Day, LLC

Published by:
Trine Day LLC
PO Box 577
Walterville, OR 97489
1-800-556-2012
www.TrineDay.com
publisher@TrineDay.net

Library of Congress Control Number: 2013951704

Estulin, Daniel
TransEvolution: The Coming Age of Human Deconstruction—1st ed.
p. cm.
Includes references.
Epud (ISBN-13) 978-1-937584-78-8
Mobi (ISBN-13) 978-1-937584-79-5
Print (ISBN-13) 978-1-937584-77-1
1. Transevolution. 2. Transhumanism. 3. Bilderberg Group. I. Estulin, Daniel. II. Title

First Edition
10 9 8 7 6 5 4 3 2 1

Printed in the USA
Distribution to the Trade by:
Independent Publishers Group (IPG)
814 North Franklin Street
Chicago, Illinois 60610
312.337.0747
www.ipgbook.com

*We are what we think.
All that we are arises with our thoughts.
With our thoughts, we make the world.*

— Gautama Buddha

To Lorena. The portrait is magical and tender, but it also remains tuned to a myth and to our expectations.

Table of Contents

Quote iii
Dedication iv
Prologue 1

1) The Economy 13

2) Genetically Modified Conspiracy 37

3) Programming the Masses 79

4) Space Exploration 115

5) Transhumanism 157

Epilogue 223
Index 225

Prologue

The year is 2015. It is a time of great innovation and technological advancement. It is also a time of chaos and conspiracies: a time of financial meltdown across the world, a time of mass population displacement, and a time when the rich are becoming immensely richer and more powerful and more fearful by the day. In 2015, corporations have more power than any government on the planet. These corporations have bankrupted the governments and have made them subservient to the interests of the moneyed elite. The final merging of One World Company Inc. is finally upon us.

"The Golden Age of cheap energy has passed. Competition for energy supplies"[1] dominates the economic landscape. Environmental degradation, the intensification of agriculture, and pace of urbanization has reduced the fertility of arable land. Food insecurity is causing mass migration on a global scale. There is severe water scarcity in some of the world's most densely populated areas – India, China and Pakistan, provoking disputes in already volatile regions that is triggering military action and large scale population movements.

Humanity is in peril. Change is inevitable. It's not the end of the world, but you can see it from here. You can feel it in the air and touch it with your fingertips.

What does our future look like? Tomorrow? In one year? In one generation? Several top-secret government studies, both in the United States and the United Kingdom, predict an eerie future: Over the next 25 years, the world will turn into a Big-Brother nightmare where small, rich elite thrives on the backs of a starving population.

"The concepts of democracy and freedom will have disappeared only to be replaced by a high-tech dictatorship based on surveillance, monitoring, mass-media indoctrination, police oppression and a radical division of social classes. The vast majority of the citizens live in third-world country conditions and are constantly subjected"[2] to poverty, famine, sickness and extermination.

By 2015, the elite see the opportunity to advance mankind towards a new Dark Age by taking the reins of Darwin's perversely racialist "survival of the fittest" natural selection evolution theory and applied social principles to develop social Darwinism.

To most, this may read like a plot from a dystopian Hollywood science fiction film, but I assure you this is real. It is all around us.

These are the conclusions of a report prepared by the British government. In December 2006, UK's Ministry of Defense prepared a secret source document on the future of humanity. The report itself was based on top-secret negotiations that took place at the conclusion of the 2005 Bilderberg Conference at the Dorint Sofitel Überfahrt hotel in Rottach-Egern, Germany. This group was handpicked by the Bilderberg steering committee in January 2005; long before the Bilderberg conference took place during three and a half days beginning on May 5.

After most of the attendees had left on the afternoon of May 8, a small, select group of Bilderbergers had retired to an exclusive Ringberg Castle, overlooking the Tegernsee in the foothills of the Bavarian Alps. The castle was a creation of Duke Luitpold in Bavaria, a member of the Wittelsbach family who ruled Bavaria for over 800 years.

Ringberg Castle

Prologue

Five months before, at a January 2005 Bilderberg pre-selection meeting, national representatives of the Bilderberg group were asked in absolute secrecy to put together a working group and prepare detailed reports on population, on natural resources availability, on conflict prevention, on economic issues, etc. The conclusions of the report, along with the conclusions of the Bilderberg meeting in May would be further discussed during the secret May 9 and 10 meetings at the Ringberg Castle.

Nobody, except the Bilderberg steering committee and a select group of Bilderberg´s most powerful members knew that this post-Bilderberg meeting even took place. The meeting would define the future of humanity and its objectives were no less than Promethean.

The people selected, were the cream of the crop of the Bilderberg elite: its longtime Chairman, Etienne Davignon, Vice Chairman of Suez-Tractebe; Francisco Pinto Balsemão, Former Prime Minister of Portugal and one of the most influential behind the scenes operators on the supranational level; David Rockefeller, a man who needs no introduction; .Timothy F. Geithner, at the time President of the Federal Reserve Bank of New York and later Treasury Secretary in the first Obama administration; Richard N. Haass, President of the powerful US think tank Council on Foreign Relations; Victor Halberstadt, Professor of Economics at Leiden University and Bilderberg´s former Chairman; Allan B. Hubbard, Assistant to the President Bush for Economic Policy and Director of the National Economic Council; James L. Jones Supreme Allied Commander Europe SHAPE; Henry Kissinger; Henry R. Kravis, Founding Partner Kohlberg Kravis Roberts & Co. and his wife Marie-Josée Kravis Senior Fellow at Hudson Institute; Queen Beatrix of the Netherlands; Matias Rodriguez Inciarte, Executive Vice Chairman Grupo Santander representing the Botin family interests; Peter D. Sutherland, Chairman, Goldman Sachs and British Petroleum; Jean-Claude Trichet, Governor of the European Central Bank; Jacob Wallenberg, representative of Sweden's most powerful family; James D. Wolfensohn, President of The World Bank and Paul Wolfowitz, at the time President designate of The World Bank.

These men and women were planning no more and no less, the future of mankind. What they decided behind closed doors of the Ringberg Castle would two years later become the backbone of the most far-sighted report in human history.

Using the British government as a once removed, deniable cutout, *Strategic Trends 2007-2036*, the 91-page report is a blueprint for UK's future strategic national requirements through the analysis of key risks and future shocks to the world's financial, economic, political, demographic and technological areas and markets. The principal output of the report focused on cross-dimensional analysis of the future context for defense over the period of one generation.

The *Strategic Trends* report is the backbone of the United Kingdom's Defense policy. The future, according to the report, "is characterized by a bewildering number of interrelated variables."[3]

By 2015, over 50% of the world's population will be living in urban rather than rural environments. The report states, "There will be a substantial growth in shanty towns and unplanned, random urban settlement, increasing the resource cost and environmental impact."[4]

Poor housing, weak infrastructure, visible marginalization, social deprivation, differential levels of poverty and a sense of grievance will increase in significance and become major political issues, "based around transnational moral justice agendas, including violent activism of varying intensity and impact."[5]

The report states explicitly that:

> In all but the most affluent societies, most of humanity will continue to experience hardship… and absolute poverty will remain a global challenge.[6]

The report goes beyond identifying the potential future military threats, and looks at the developments in areas that will shape the wider strategic context within which Defense will have to interact. Key theme of the report addresses population and resources of the planet Earth. In stark terms, it outlines:

> … an increased risk of humanitarian catastrophe caused by a mixture of climate change, resource pressures, uneven distribution of wealth, the effect of disease and the failure of authorities to cope with population growth and urbanization.[7]

In one generation, between 2007 and 2036, world's population explosion will go from 7 billion people to almost 10 billion with less developed countries accounting for 98% of world population growth. By 2036, nearly two-thirds of the world's population will be living in areas of water-stress. The lack of food, water, medicine, proper hygiene, education and basic human necessities could spell – collapse.

Without mincing words, the report states explicitly:

> … growing gap between majority and a small number of highly visible super-rich is likely to pose an increasing threat to social order and stability. Faced by these challenges, the world's under-privileged might unite, using access to knowledge, resources and skills to shape transnational processes in their own class interest.[8]

The result of the growing desperation on the part of humanity will result in "civil war, intercommunal violence, insurgency, pervasive criminality and widespread disorder."[9]

Faced with the financial markets on a downward spiral to Hell and a global economic collapse, the report predicts "a severe pricing shock, possibly caused by an energy spike or a series of harvest failures,"[10] which would "trigger a domino effect involving the collapse of key international markets across a range of sectors."[11] The impacts of this collapse, transmitted throughout the globalized economy, could result in a breakdown of the international political system and complete economic collapse.

Does that mean all of humanity is in peril? No, it does not. Because even with the "erosion of Civil Liberties," the super-rich will be protected through "technological advancements and pervasive surveillance." Coupled with "intrusive, highly responsive and accessible data-bases, the emergence of a so-called 'surveillance society' will increasingly challenge assumptions about privacy, with corresponding impacts on civil liberties and human rights."[12]

With the destruction on nation-state republics and the creation of mega economic blocks linked to each other through a globalized marketplace, independent countries will be "replaced by Mega Cities." The report defines a Mega City as a "large city in a developing region" with a population base of over 20 million people. Caused by a massive population displacement, cities will swell to unimaginable proportions "which will already have experienced endemic lawlessness and high levels of violence."[13]

Unable to cope with an influx of peoples, the Mega Cities:

> ... will fail before 2035. The effects will be equivalent ... to state failure, which city failure may, in turn, precipitate.[14]

Based on recent experiences in the United States, the military stabilization of a major city could demand Martial Law, or as the report deceptively states: "... a comprehensive Inter-Agency approach, specialist skills, and an enduring operational commitment."[15]

Technological advancement and supremacy will require that wars be fought not state-on-state basis, but rather as an "endemic Urban-Based, Irregular Conflict against adversaries with highly-developed urban survival and combat skills."[16] These are called *social turbulences*.

Social Turbulence

A theory of social turbulence, a so-called "softening up effect of future shocks" was developed by two prominent psychologists, Eric Trist and Frederick Emery – wherein a population could be softened up through mass phenomena such as energy shortages, economic and financial collapse, or terrorist attack. "If the 'shocks' were to come close enough to each other and if they were delivered with increasing intensity, then it was possible to drive the entire society into a state of mass psychosis," claimed Trist and Emery. They also stated that "individuals would become disassociated, as they tried to flee from the terror of the shocking, emerging reality; people would withdraw into a state of denial, retreating into popular entertainments and diversions, while being prone to outbursts of rage."

How does the *Strategic Trends* report deal with social turbulence, mass psychosis and the softening up of the population?

> Increasingly, regular military forces will deploy in environments where armed irregular forces, for example armed followings, gangs, bandits, semi-official militias, Private Military Companies (PMCs), terrorists and insurgents, are operating, often as adversaries, but sometimes as neutrals or even as partners.[17]

This is called breaking the morale through the strategy of terror. In fact, we are talking about two sides of the same coin here. On one side, guiding the covert, subtle manipulation and control of thought and human consciousness through the power of television in particular and entertainment in general; "employed on an ever growing list of those deemed as enemies of America as well as on a confused and agitated American public – whose corporate news networks frame and manage an increasingly shallow narrative while engaging in a kind of Orwellian Kabuki Theatre of fairness and balance;"[18] while "on the other side, directly and overtly shifting the paradigm, changing the basic concepts, widening the parameters, and changing the playing field and all the rules of play by which society defines itself within an exceptionally short period of time."[19]

One of the key individuals involved in psychological warfare against the population through manufactured social turbulence is Kurt Lewin, a pioneer in-group dynamics who was part of the early Frankfurt School and fled Germany when Hitler took power. This passage from his book *Time Perspective and Morale*, shows his understanding of psychological warfare:

> One of the main techniques for breaking morale through a 'strategy of terror' consists in exactly this tactic – keep the person hazy as to where he stands and just what he may expect. If in addition frequent vacillations between severe disciplinary measures and promises of good treatment together with

spreading of contradictory news, make the 'cognitive structure' of this situation utterly unclear, then the individual may cease to even know when a particular plan would lead toward or away from his goal. Under these conditions even those who have definite goals and are ready to take risks, will be paralyzed by severe inner conflicts in regard to what to do.[20]

Over the past 50 years, research in the fields of psychology, sociology and psychiatry has shown that there are clearly marked limits to the amount of changes, and the nature of them that the mind can deal with. According to Science Policy Research Unit [SPRU] at Tavistock's Sussex University facility, "future shocks" is defined "as physical and psychological distress arising from the excess load on the decision-making mechanism of the human mind." In other words:

> … a series of events, which come so fast that the human brain cannot absorb the information."[21] One scenario is called superficiality. After continuous shocks, according to Emery and Trist, the large targeted population group discovers that it does not want to make choices any more, reducing the "value of his intentions…. This strategy can only be pursued by denying the deeper roots of humanity that bind … people together on a personal level by denying their individual psyche.[22]

Apathy takes over, often preceded by mindless violence such as is characteristic of the Los Angeles street gangs in the 1960s and the 1980s, what Emery and Trist call organized social response to dissociation, as described in the pages of Anthony Burgess' novel *A Clockwork Orange*, a society dominated by infantile animal like rage. "Such a group becomes easy to control and will docilely follow orders without rebelling, which is the object of the exercise," add Trist and Emery. What's more, the dissociated adults cannot exert moral authority over their children, because they are too involved with their own infantile fantasies, brought to them through their television set. And if you doubt what I am saying, look at the older generation today as they have accepted the moral decadence of the no-future generation of its children, rather than seek conflict, and in the process, the adults have come to accept a lower moral standard.

Just as in Aldous Huxley's drug-controlled *Brave New World*, there are no moral or emotional choices to make here, the "flower children" and the drug soaked rebellion of the Vietnam era is a perfect example of how this scenario functions.

These "frequent vacillations" pass through several scenarios: "Stable, at which point, people more or less are able to adapt to what is happening to them, or it is turbulent, at which point people either take actions to relieve the tension, or they adapt to accept tension-filled environment. If the turbulence does not cease, or it is intensified, then at a certain point people cease being able to adapt in a positive way. According to Trist and Emery, people become maladaptive – they choose a response to tension that degrades their lives. They start to repress reality, denying its existence, and constructing increasingly more infantile fantasies that enable them to cope. Under the conditions of increasing social turbulence, people change their values, yielding to new degraded values, values that are less human and more animal-like."[23]

The second scenario is "segmentation of society into smaller parts. In this scenario, it is every group, ethnic, racial, and sexual against the other. Nations break apart into regional groups, those smaller areas in turn fissure into even smaller areas, along ethnic lines."[24] Trist and Emery refer to it as "enhancement of in-group and out-group prejudices as people seek to simplify their choices. The natural lines of social divisions emerge to become barricades."

The *Strategic Trends* report has an answer for that as well. Increasingly non-state actors:

> … will be wielded by a broader spectrum of individuals and agencies, even by criminal, terrorist and insurgency groups as a means of complementing their more coercive, violent activities. These groupings will be highly volatile, dissolving either when interests are achieved, or when more favorable opportunities present themselves, with those least constrained by legal accountability or moral considerations the most likely to be effective at manipulating the use of Soft Power.[25]

Society's response to such a psychological and political disintegration is the Orwellian fascist state, modelled on George Orwell's book *1984*. In the book, "Big Brother" regulates the lives and conflicts of people within a society; a never ending conflict:

> ... is waged by each ruling group against its own subjects, and the object of the war is not to make or prevent conquest of territory, but to keep the structure of society intact.[26]

This never-ending conflict is foreseen in *Strategic Trends* report.

> Information Communications Technology, ICT, is likely to be so pervasive that people are permanently connected to a network or two-way data stream with inherent challenges to civil liberties; being disconnected could be considered suspicious.[27]

The growing pervasiveness of ICT will enable communities of interest to be established rapidly and for the quick and coordinated mobilization of significant numbers of people.

> Rapid mobilization – 'Flash mobs' – may be undertaken by states, terrorists and criminals, and may involve dispersed communities across international boundaries, challenging security forces to match this potential agility and ability to concentrate.[28]

Crucially, "this level of sophistication will require a concerted, comprehensive application of all the instruments and agencies of state power, together with cooperation from all relevant authorities and organizations involved in settling a crisis or resolving a conflict."[29] What does it mean? It means, Martial Law.

With resistance broken, the third "future shocks" scenario is the most intense, involving a withdrawal and retreat into "private world and a withdrawal from social bonds that might entail being drawn into the affairs of others."[30] Trist and Emery are convinced that men will be willing to accept "the perverse inhumanity of man that characterised Nazism." Not necessarily the structure of the Nazi state, but the moral outlook of Nazi society.

To survive in such a state, people will either have to submit to the state or go underground. Again according to the *Strategic Trends* report,

> All likely future opponents will have recognized the advantages of going underground if they wish to avoid the surveillance, targeting and penetrative capabilities of sophisticated military forces, particularly those deploying air platforms and systems. In future, states will seek to site most of their major nodes and the majority of their decisive fighting power underground. Similarly, irregular opponents will base themselves in underground networks, both for offence and defense, especially in complex urban spaces.[31]

Techniques for overcoming resistance, comprehensive application of all the instruments and agencies of state power, rapid mobilization, "Flashmobs," pervasive Information Communications Technology, segmentation of society, endemic lawlessness and high levels of violence, intrusive, highly responsive and accessible data-bases, the emergence of "surveillance society," erosion of Civil Liberties, humanitarian catastrophe.

This is what the near future looks like. We instinctively reject such conclusions, even if they are supported by solid evidence in a Bilderberg-authored and government-prepared secret report. "However, we must have the courage to let the evidence tell the story, because, as the reader will learn, what is at stake is not simply our view of reality. Whether or not our worldview is challenged, or shattered, by the revelations in this material, doesn't change the truth of the matter, which is that the conspiracy is real, that it is ongoing – and that September 11, 2001, was just a prelude to what they have planned for us."[32]

We are truly at the gates of Hell. And the roads we take will now terminate whether we will live in the XXI century as nation state republics or subjugated, culled and dehumanized crop of slaves.

Endnotes

1   http://www.resilience.org/stories/2007-02-27/dcdc-global-strategic-trends-programme-2007-2036.
2   Ibid.
3   DCDC *Strategic Trends Report*, p.v.
4   Ibid, p. 9.

5 Ibid, p. 13.
6 Ibid, p. 33.
7 Ibid, p. 6.
8 Ibid, p. 80.
9 Ibid, p. 68.
10 Ibid, p. 78.
11 Ibid.
12 Ibid, p. 61.
13 Ibid, p. xii.
14 Ibid, p. 28.
15 Ibid, p. 30.
16 Ibid.
17 Ibid, p. 70.
18 Elizabeth Gould and Paul Fitzgerald, "9-11, Psychological Warfare and the American Narrative," http://www.boilingfrogspost.com/2011/08/10/911-psychological-warfare-the-american-narrative-part-i/.
19 John Quinn, *NewsHawk*, 10 October, 1999.
20 K. Lewin (1942), "Time Perspective and Morale," in G. Watson, ed., *Civilian Morale*, second yearbook of the SPSSL, Boston: Houghton Mifflin.
21 John Coleman, *The Story of the Committee of 300*, Global Review Publications, 4th edition, 2006.
22 Ibid.
23 Lonnie Wolfe, Turn off your TV, *New Federalist*, p. 12-13, 1997.
24 Ibid.
25 DCDC *Strategic Trends Report*, p. 55.
26 George Orwell, *1984*, Signet 1961, 270p.
27 DCDC *Strategic Trends Report*, p. 58.
28 Ibid, p. 63.
29 Ibid, p. 70.
30 Lonnie Wolfe, Turn off your TV, *New Federalist*, p. 13, 1997.
31 DCDC *Strategic Trends Report*, p. 73.
32 S.K. Bain, *The Most Dangerous Book in the World*, TrineDay, 2012.

Chapter One

The Economy

The delusions we have about the economy arise from the fact that some think that economy is a matter of money. However, money is not a determinant of wealth. The statistical measures of money flows have nothing to do with the actual forecasting of wealth. What affects the planet is the development of the mind of the individual.

It is not money, it is not statistics, and it is *not* monetary theory that determines the way an economy works. It is physical. And physical includes the fact that the human brain, which is unlike any other brain. No animal can file a patent – only human beings can do that. That is one of the ways in which we organize our social system – by adopting social customs, as to how we behave as people, as human beings. And, if we have defective customs, we are going to have problems – and these problems are *foreseeable*.

Dear reader, I want you to understand that what we are witnessing in the world today – *the wholesale destruction of the world's economy* – is not an accident, nor is it a miscalculation or the result of political shenanigans. This is being done on purpose, *absolutely on purpose*. The reason is because our current corporate empire knows that "progress of humanity" means their imminent demise.

Empires cannot survive in a world where there is widespread technological and scientific progress. An imperium requires a world where people are dumb and servile like sheep. Empires seek to destroy the very structures, such as nation-states, which support meaningful life and human progress. The "powers-that-be" deliberately target these nation-states (independent countries), and their economies – to destroy the nations, and thereby maintaining their own imperious power. *This is by design.*

Empires are not a King or a Queen sitting on a gold plated throne. Empires are above Kings. *It is a system of control* – total control. The control of everything by an international monetary system that is itself run by the international banks.

Globalization, you see, is nothing but a new form of empire, with the goals: the elimination of the nation-state, the elimination of liberty, and the elimination of rights.

Now, I know many people will comment: "What empire? Empires ceased to exist a long time ago."

Our current imperium is a complex system of central banks, investment banks, hedge funds, private equity funds, insurance companies, and shadow banking. A financial system that has been responsible for a paradigm-shift in the last half a century, away from the production of physical goods and real economy, and more and more to speculation, to the idea of profit maximization, and pure monetarism. That's todays' Empire!

How Money Works

Like I said before, the economy has nothing to do with money. What the elite have created is an Empire – with them in control. Too many believe that is all about money. But money is simply an instrument. Money, in itself, doesn't affect the development of the planet. There is a false belief that there is an intrinsic value expressed by money, but the value expressed is not as a quantity per se. It is only the relative effects of its increase, and/or decrease of its physical potential relative to the population-density of a society. The value of money lies not in its individual exchange, but in its function as a unifying dynamic of the social processes of a nation.

What affects the development of our planet? The human mind. That is how mankind is measured. What separates us from animals is our ability to discover universal physical principles, which then increase and improve humankind. Our minds allow us to innovate, and subsequently improve the lives of our populations. The development of mankind, the development of the power of individuals and the nations, *depends* upon scientific development, scientific discovery and technological progress.

By cutting down productivity, by disrupting investment in infrastructure, and by forestalling inventions and technology, the elite wish to force a population reduction. Because if you keep people stupid, and not too numerous, a minority may excercise near total control.

Today's monetary crisis reflects the insanity that has been imposed during the process of destroying the physical economy. The reason we are having this "crisis" is not because of fluctuation in the financial markets. But because we are in a hyperinflationary mode: if you take the total amount of money purported to be in circulation, and you take the percentage of that money which has a correspondence to physical reality, and you find it is diminished – almost to zero. But that is not the problem! It is our per capita physical production, and the fact, that the resources on which production depends, have collapsed.

The future of mankind comes from our physical economy, and the physical transformations that we make from the world that surrounds us. Let me give you an example. You take a pile of dirt, which is rich in iron ore, and you refine the ore, and you use that ore to make iron, and then you use that iron to make steel, and you use the steel to make machine tools that then allow you to build all sorts of other things: automobiles, trains, space shuttles, nuclear reactors. You have forged ahead.

At each step of the process the transformation you get has greater value to society, is worth more in physical terms than what went into it. The output is more than the sum of the inputs.

You see, when we build infrastructure, we are actually reorganizing the physical space-time of the planet, allowing our world to attain higher and higher levels of efficiency and beauty.

So, if your system is increasingly moving to higher levels of efficiency, such as from a coal based economy to an oil based economy to a nuclear based economy, there would be an increase in the productive power of human labor at every step of the way –*that's the way that actual wealth is created*. People doing productive things. Money is simply the way of facilitating trade among people. It possesses no magical powers. No intrinsic value.

Why Is The Empire Destroying The Financial System?

There are currently seven billion people on Planet Earth, a small blue orb meandering through space with limited natural resources and an ever expanding population. Food and water are becoming ever scarcer.

For example, according to a study by NASA, Gravity Recovery and Climate Experiment, India's water tables have fallen by about one-fifth over the past two years.[1] India's agricultural sector currently uses about 90% of its total water resources. As water scarcity becomes a bigger and bigger problem, India could be faced with catastrophic food shortages.

Soon India may be forced to import more food from abroad. With a population of 1.2 billion people, the tiny food surplus that exists in the world today would vanish immediately, creating an unsustainable situation for the world at large.

The elite understand this. David Rockefeller understands this. The Kings and the Queens of the world understand: a larger population equals fewer natural resources and more food and water supply shortages.

In fact, the elite, the oligarchic controllers of empire throughout Europe, understood this by the middle of the eighteenth century: that levels of technological progress and development are directly proportional to population growth. Without scientific and technological progress, a level of population increase cannot be sustained because with an ever-expanding population, you need more and more sophisticated technology to sustain the populace.

And with technological progress, oligarchies are not tolerated. Nations, which foster the creative-mental development of their populations, produce a people that will not tolerate oligarchic forms of rule indefinitely.

Therefore, from the point of view of an oligarchical elite, if they want to completely control a planet, they must reduce the population to a more "manageable" number. Remember, seven billion people and growing is a lot of mouths to feed. This is something that Rockefeller and company understands, even if we don't. For the elite to eat, you and I have to die. How is that for a solution?

Depopulation & Club of Rome's "Limits to Growth"

The most important institution in the world to push a Malthusian depopulation scheme is the Club of Rome. Its members are some of the leading citizens of the planet: David Rockefeller, Michael Gorbachev, the King and Queen of Spain, Princess Beatrix of the Netherlands, King Philippe of Belgium.

Founded in April 1968 by Alexander King and Aurelio Peccei, the Club of Rome, mainly consists of members of Venetian Black Nobility, descendants of the richest and most ancient of all European families. They controlled and ran Genoa and Venice in the XII century.

In 1972, the Club published one of the most pernicious documents of all time, *The Limits to Growth*. The report declared that Earth was going to run out of its limited resources during the next 40 years. Therefore, according to this account, in order for mankind to survive, we would have to adjust our lifestyle, and numbers, accordingly.

According to the Club of Rome, in order to survive, mankind must reduce its dependence on technology; roll back its drive for progress, technological innovation and advancement, and impose a worldwide regime of economic *controlled disintegration*.

> Since that time, the *Limits to Growth* thesis has been inserted within government and supranational government institutions worldwide, so-called educational institutions, university curricula, and much more – basically every aspect of popular culture. The results have been the total deindustrialization, wars, and genocide we see today.[2]

The end result is the collapse of the world economy, and even with their version of "unlimited" resources, which includes no breakthroughs in science, or the development of any revolutionary new technologies. If you break through the Babylonian verbal confusion, a report by the Club of Rome leaves little room for doubt as to their real agenda: "In search for a new enemy to unite us, we came up with the idea that pollution, the threat of global warming, water shortages, famine and the like would fit the bill." The report concluded with the following, "The real enemy, then, is humanity itself."

Thus, a leading international institution is pushing policies of retrogression in technology, and the reduction of the world's population by several billion people – that's *genocide* – in case you didn't know it.

However, before they can reduce population and tame the herd, *they must destroy the world's economy by destroying consumer demand*, which is accomplished by making us all poorer.

You may wish to know why Rockefeller and company would want to destroy demand. Wouldn't they suffer financially as well? The answer is *no*. They will not suffer at all, because they already control most of the world's wealth. Their primary concern at this juncture in history is to ensure the survival of their *breed*. And once more, for them to survive in an age of depleting natural resources, *most of us must die*.

* * *

We have all heard of the Great Depression. However, many of us haven't been taught our own history and do not understand what happened during that period.

Contrary to what "official" history books tell you, the Great Depression was not an event that wiped out U.S. capitalists. It was an event that made the rich richer, transferring the wealth of the people into the hands of an already wealthy elite.

The Bank of America made billions through real estate foreclosures from 1929-37. Don't believe for a minute that the richest of the rich will be hurt by any economic collapse. The only ones hurt will be you and me.

A question arises, how do they destroy demand? By destroying the world's economy on purpose. In other words, "controlled disintegration." Which was precisely the cornerstone of another policy paper prepared by another elite group: the Council on Foreign Relation's *Project 1980s*. Controlled economic disintegration and the dismantling of the globe's advanced scientific industrial concentrations were a major component of that report. CFR, one of the oligarchy's central institutions in the United States, called this project, "the largest undertaking in its history."

The 33-volume CFR report constituted blueprints, which the oligarchy used its power to institute during the second half of the 1970s and the 1980s. They imposed one of the most profound shifts in economic and nation-state policy during the 20th century – the paradigm shift to a post-industrial economy.[3]

What does "controlled disintegration" mean? The world economy would be pushed into collapse – but not in an haphazard fashion. With the oligarchy controlling the disintergration process, it would be necessary to deliver economic shocks to carry this out: oil price hikes, credit cutoffs, interest rate instability. Forcing the world economy to zero, and eventually negative growth rates.

Simultaneously, there was the creation of the spot market in oil, of the euro bond markets, the derivatives market, and the expansion of the offshore banking apparatus as well as the laundering of large quantities of drug money through some of the world's largest banks.

Over the past several years, some of the world's largest banking institutions have been caught laundering billions of dollars in illegal proceeds from the drug trade through their accounts: Wachovia Bank, HSBC, Citigroup, and even Coutts, the Queen of England's private bank.

Behind this entire initiative there appears the Inter-Alpha Group of banks. The Inter-Alpha Group, since its founding in 1971 as a European banking cartel by Lord Jacob Rothschild, has been an epicenter of operations – all directed from London. The Inter-Alpha Group includes financial heavyweights such as the Royal Bank of Scotland, the Portuguese Banco Espirito Santo, the Spanish Banco Santander, the Dutch ING, the French bank Société Générale and Germany's Commerzbank, among others.

The group was actually created in the later stages of World War II. Jacob Rothschild directed it from London. The Rothschild banking network, since its inception in Frankfurt in late 1700, has been an operation of the Black Nobility, a global financial conglomerate conspiracy that dates back to the Fourth Crusade.

One of Rothchild's first sponsors was the Thurn und Taxis family of Bavaria, controllers of the Venetian intelligence and one of the principal families of the Habsburg Empire of Austria.

This Venice connection is the true source of financial power of the legendary Rothschild family.

Despite the enormous wealth that these banks represent by themselves, they did not have all the funds necessary to transform the world according to their plan. They provided the initial capital, and then used their power of controlling the money of others, to create the markets and the institutions used to control world markets. This group originated the hedge funds, the private equity funds and other domineering financial tools – the dark side of the Inter-Alpha Group.

To achieve its objective, they first built a banking unit in post-war Europe World War, as the base of what appeared to be a new universal financial structure. However, it was actually a return to an imperial model that existed before the American Revolution. The planning of this new Europe began even before the fighting ceased, and soon led to the creation of the European Coal and Steel Community in 1951, and then the formation of the European Economic Community in 1957. These were initial crucial steps towards today's European Union and its supranational currency, the euro.

With these measures and their subsequent elimination of national sovereignty, this new/old economic empire began the process of building a borderless European financial system. In quick succession, came the development of the Eurobond and the Eurodollar markets, and mega-banks. These banks, mostly based in London, merged with other British banks and banks based in Europe, Asia and the Americas. These mega-banks were designed to flank national banking regulations, and as such, represented the beginning of "globalization" (i.e. imperialization) of finance.

The real power of the Inter-Alpha Group, however, is not in the individual banks themselves, but in the changes that the Inter-Alpha operation has fomented in the world economy. The Inter-Alpha project turned the global financial system into a giant casino: a theme park for the investment banks, which are the "speculative arm" of the commercial banks and hedge funds.

Trilateral Commission

Another organization closely linked to the Bilderberg Group and the Council on Foreign Relations is a little understood entity – the Trilateral Commission (TC). David Rockefeller established the organization in 1973. The people who belong to the Trilateral Commission all appear to share the same anti-nationalist philosophies, and to try to prevent the internal nationalistic forces within their respective countries from exerting influence on policy.

Setting up the TC became Rockefeller's plan to encourage:

> … unity among the industrialized powers, so that together they could achieve his goal of a more integrated global political and economic structure.[4]

Rockefeller strategy "also reveals something fundamental about wealth and power: it does not matter how much money one has; unless it is employed to capture and control those organizations that produce the ideas and the policies that guide governments and the people who eventually serve in them, the real power of a great fortune will never be realized."[5]

Despite the primarily financial nature of Trilateral Commission's motives and methods, their political goals haven't changed in forty years:

> Although the Commission's primary concern is economic, the Trilateralists have pinpointed a vital political objective: to gain control of the American Presidency.[6]

One of Trilateral Commission's most notable recruits was candidate and later President, Jimmy Carter. For the complete story on Carter's selection for President, please see *The True Story of the Bilderberg Group*.

Once the Trilateral Commission's Jimmy Carter was installed as President, the oligarchy transferred the CFR's *Project 1980s* into his administration. The top personnel of *Project 1980s* became the top leadership of, and beginning in 1977 ran the government of Jimmy Carter. Two of the Project's nine directors, W. Michael Blumenthal and Zbigniew Brzezinski, were appointed Treasury Secretary and National Security Adviser, respectively. Cyrus Vance, who headed a *Project 1980s* working group, was appointed Secretary of State. And Paul Volcker, spokesman for

Project 1980s' "controlled disintegration"[7] policy, became Federal Reserve Board chairman.

> Starting the week of Oct. 6-12, 1979, Volcker began raising interest rates through raising the federal funds rate and increasing certain categories of reserve requirements for commercial banks. He kept pushing rates upward, until, by December 1980, the prime lending rate of U.S. commercial banks reached 21.5%....
>
> The effects of this policy were swift and devastating, especially because the oligarchy had used two oil hoaxes during the 1970s, to send oil prices shooting upward. In the United States, industrial and agricultural production collapsed by huge amounts. Between 1979 and 1982, the production of the following critical U.S. manufacturing industries fell by the following amounts on a per-capita basis: metal-cutting machine tools, down 45.5%; bulldozers, down 53.2%; automobiles, down 44.3%; and steel, down 49.4%.[8]

Does that sound like a push towards a post-industrial society? You bet it does.

1973 Bilderberg Meeting and a Planned Oil Hoax

In early 1973, the dollar was falling and the French, the German and the Japanese economies were really beginning to boom. In the beginning of 1973, the West German deutschmark had already smashed the British pound and by July-August was on its way to gaining hegemony over the ailing US dollar.

In May 1973, the Bilderberg Group met at an exclusive resort at Saltsjobaden, Sweden:

> Certain elites connected with the money center banks in New York decided that it was time for a major shock to reverse the direction of the global economy, even at the cost of a recession in the American economy – that didn't concern them so much as long as they were in control of the money flows.[9]

The key point on the Bilderberg meeting agenda was the oil shock of 1973 – the 400% targeted increase in the price of OPEC oil in the near future. Economist William Engdahl explains:

The entire discussion was not how do we as some of the most powerful representatives of the world's industrial nations convince the Arab OPEC countries not to increase oil prices so dramatically. Instead they talked about what do we do with all the petrodollars that will come inevitably to London and New York banks from the Arab OPEC oil revenues.

The oil shock came two years after the free floating of the dollar, when the dollar was essentially falling like a stone, because the U.S. economy was starting to show major ruptures from the Post-World War II period when the U.S. industry was a world class leading industrial power and the gold reserves and everything else was in an ideal correlation to one another.[10]

The Real Reason Behind The 400% Oil Price Increase

The oil price jumping by 400 per cent in 1973/74 saved the dollar. The dollar floated up on a sea of oil. We need to remember that Nixon broke the link of the dollar with gold unilaterally in August of 1971, and after that time it had plunged by some 40 per cent against the major trading currencies like the Deutsche Mark and the Japanese Yen. What saved the dollar? What saved Wall Street and the power of the dollar as a financial entity? It was not the U.S. economy by any means, it was the 400 per cent OPEC price hike.

The price shock halted growth in Europe, and smashed the industrialization of the developing countries in the Third world, which had been enjoying a rapid growth dynamic in the early 1970s. The massive oil price increase tilted power into the direction of Wall Street and the dollar system.

All of that was aimed at continuing the systemic imperial process of looting the actual productive wealth of the major nations of the planet. This price hike scheme ultimately created an enormous volume of wealth transfer, nominally into the OPEC countries, the so-called petro-dollars, but the money went to London and Wall Street to be managed. Thus, the financial oligarchy, in the major centers, used the oil-price hoax to establish an absolute domination over the world's credit system, and to make sure it no longer went to civil and cultural development.

They used it to:

... fund operations to transform the United States from within, including the takeover of the U.S. banking system and the cartelization – under the euphemism of mergers and acquisitions – of corporate America. Wall Street was transformed into a giant casino, where betting on financial instruments replaced investing, and the connection to reality was severed. At the same time, the petrodollars helped fund cultural warfare operations against the American people, to keep them blind to the damage being done, or even conning them into believing it was progress.[11]

The orchestrated oil hoax of 1973-74, with its introduction of financial speculation in the oil market via the spot market, created a huge pool of petrodollars, with which the City of London could wage war against nations. These petrodollars, combined with the proceeds of the British Empire's "Dope, Inc." drug trade, were instrumental in restructuring Wall Street in the 1970s, paving the way for the junk bonds of the 1980s and the derivatives of the 1990s.[12]

Derivatives, Mortgages And The Speculative Bubble

Most of us have heard the term, "speculative bubble." What does it mean and where does it come from? Once someone makes a decision to create a bubble, basically it is a pyramid scheme. De-linking the financial gains from the re al economy, which is what you have to do if you are killing the real economy, and if you want to build up a speculative bubble, you have to divorce it from reality, and derivatives are a way to do that.

It's like creating a game at a casino table. Derivatives are side bets on the movements of various things like bonds, the value of bonds, interest rates, and currency rates. You speculate on all these things, and you can bet on those speculations.

The End of the Line

As the speculative bubble came to dominate the U.S. and world economies, feeding it became paramount. Among other things, this led to a sharp run-up in real estate values, to provide 'wealth' which could be turned into mortgage debt, and then into a wild assortment of securities to be used, with lots of leverage, to play in the derivatives markets. To keep the mortgage-debt flowing, as prices rose into the stratosphere, the bankers repeatedly loosened the requirements for home loans. This process, which was driven by the banks and the derivatives market, ultimately exploded. This was falsely portrayed as a 'subprime' crisis, but in reality it was the death throes of the financial system itself.

In mid-2007, the failure of two Bear Stearns hedge funds signaled the collapse of the global securities market, as speculators realized the game was over and began to try to cash out. The market for speculative paper quickly dried up, sending the nominal valuations plunging. The market, which had grown phenomenally through leverage, began to collapse in a reverse-leverage implosion. Speculators had borrowed trillions of dollars to place bets, gambling that they would win enough to pay back their loans and still turn a nice profit. This game worked for quite a while, but it quickly turned nasty when the market seized up. Suddenly, the speculators found themselves losing on their bets, leaving no profits to pay off their loans, and thus losing on both ends. Assets began vaporizing by the trillions, and worried lenders began demanding more collateral on margin calls, causing sales of assets that further depressed prices, in a vicious, reverse-leverage spiral.

The 'solution' to this blowout adopted by the central banks, was to begin to flood the financial markets with liquidity, through a series of interest rate cuts and cash injections. Though they had sworn to impose discipline on the markets, the central banks quickly capitulated under the pressure of enormous losses, in a hyperinflationary panic. The injections quickly escalated from the billions, to the tens of billions, to the hundreds of billions, as they raced to plug the holes caused by the savage deflation of the valuations in the system. But no matter how much money they injected, the system kept collapsing. The money pumped into the bailout – money that serves no economically useful purpose – will only accelerate the process. This means that the faster the government pumps in the money, the faster the value of the dollar will collapse, and the faster the global economy will collapse.[13]

Mergers & Acquisitions – The World Company Inc.

Another term you are familiar with is "mergers & acquisitions." Mergers and acquisitions is a euphemism. You find many euphemisms for imperialization because imperialization is a bad word. In 1968 at the Bilderberg conference in Canada, George Ball, a senior managing director for Lehman Brothers as well as the Undersecretary of State for Economic Affairs with JFK and President Lyndon B. Johnson, announced a project to build what he called the "World Company."

The idea world globalists like to promote is that nation-states are outmoded and an archaic form of government, and that in a Malthusian world they can't be relied on to address modern needs of society.

> For Ball, the very structure of the nation state, and the idea of commonwealth, or of a general welfare of a people, represented the main obstacle against any attempt of freely looting the planet, and represented the most important impediment to the creation of a neo-colonial world empire.[14]

In other words, according to Ball and others in the Bilderberg Group, the resources of any country do not belong to that country, but to the World Company Incorporated run by the Elite.

And so what is needed is a new form of government, one that will more *freely* distribute the world's resources. And that new form of government they decided is the *corporation*.

What George Ball called the *World Company* could then become a new government that would greatly surpass, *in authority and power*, any government on the planet.

What we have seen from this cabal has been the gradual collapse of the US economy, beginning in the 1980s.

> The corporate raiders, financed by the dirty-money junk bond network, bought up significant chunks of corporate America, and terrified the rest. The raiders' targets and those who feared they might become targets, turned to Wall Street's investment banks and law firms for 'protection.' As such, the leveraged buyout/junk bond operation functioned as a giant protection racket, destroying some as a way of collecting tribute from the rest. At the same time, dirty money

poured into the real estate market, notably through the giant Canadian developers… These firms built the skyscrapers, which were then filled up with service workers – bankers, lawyers, accountants, clerks, and other white-collar types…

The pouring of hot money into the real estate markets caused real estate prices to rise. The 'wealth' created by these rising values provided more money to pump into the bubble…The speculator went from being the enemy to being the role model…The old-style productive industry became the realm of 'losers,' replaced by the hot new 'industries' of finance and information…

The effect of all this deregulation and speculation has been the decimation of the physical economy of the United States. Over the last three decades, the productive capacity of the US economy has been cut in half, measured in terms of market baskets of goods on a per-capita, per-household, and per-square-kilometer basis. At the same time, the monetary claims on that declining production have risen hyperbolically.[15]

Much of the control of the World Company Incorporated is not on the surface.

It is exercised through the London Stock Exchange, through the London International Financial Futures Exchange, through the London Metal Exchange, and the International Petroleum Exchange. These are the institutions of World Company Ltd where the actual disposition of the physical assets being traded is determined, not to mention the layers upon layers of speculative financial instruments created, that are now in full collapse, and threatening to bring the physical economy of the world down with it.[16]

And if you look at globalization, that's exactly what this is. Beginning late 1960s into the 1970s and 1980s, the USA and the rest of the world was taken over by this rash of mergers, an ever-larger consolidation of industrial companies, of agricultural companies, of financial companies. They were slowly building these giant cartels to the point where we see now, today, giant cartels which control the resources of the world – *effectively running the world.*

Bankers are running the corporations, and their cartels. These cartels control the necessities of life, and they are now more powerful than nations. This whole World Company project, in a sense, is a return to the old monopolistic days of the British East India Company with a modern computerized face. What should frighten people most is: the elite have actually done what they announced they would do, way back in 1968. Doesn't that scare you?

Let me give you an example of what One World Company means. Take Royal Dutch Shell (RDS), this mega-corporation is a product of union between the largest British and Dutch petroleum interests.

Royal Dutch Shell's World War II chief, Henri Deterding was a notorious backer of Adolf Hitler. RDS's banker, Lazard created Banque Worms out of a Shell-connected transport company. Banque Worms was seen as a notorious Vichy regime supporter and financial backer. Furthermore, Royal Dutch Shell has funded cultural warfare operations against the United States, and the rest of the world, the so-called "dumbing down of society," including the creation and subjugation of the environmentalist movement, which is given a "role" in the post-industrial gang. *Playing on people's hopes for a better world.*

RDS shares an interlocking directorship with the Dutch ING bank; Dutch chemical company Akzo Nobel; Unilever, an Anglo-Dutch group that controls large chunks of world's food production; and Rio Tinto, which along with Anglo-American control from 10 to 24% of the Western world's minerals output. RDS also shares directorship with Boeing, Lloyds Bank, UBS and AXA (one of the world's largest insurance companies).

The British part of the Royal Dutch Shell is BP, British Petroleum. BP, shares an interlocking directorship with Royal Bank of Scotland, and HSBC (which was recently caught laundering billion of dollars of Mexican drug money), Akzo Nobel, Unilever, Roche pharmaceutical, and Goldman Sachs (whose former employee Mario Monti was the one time unelected caretaker Prime Minister of Italy), Rolls Royce, General Electric, Bank of America, Lloyds Bank, KPMG and GlaxoSmithKline pharmaceutical company.

On the next level down, many of these companies are further interlocked amongst themselves. For example, HSBC is interlocked with BP as well as Shell, gold producer Anglo-American, *Financial Times*, one of the leading business papers in the world, the *Economist*, Imperial Chemical Industries, GlaxoSmithKline, Rolls Royce and Kleinwort Trust through a major German investment bank Dresden Kleinwort Benson.

Furthermore, each and every one of the above mentioned corporations, are interlocked amongst themselves, creating an unbreakable, self-perpetuating system, a virtual spider-web of interlocked financial, economic and industry interests with the World Company model at the center. A fondi system.

2009 Bank Bailout

Do you remember the 2009 bank bailout? What was the real reason why the bankrupt banking sector was bailed out? Were Wall Street and the US Government truly thinking of "saving America" as they said? Or was there another reason to it?

In fact, hidden from the public, there was a far more sinister reason for the bailout. Please understand the bailout was a massive fraud. Under the guise of saving the economy, the bankers transferred huge amounts of debts from private hands, from the banks and other powerful interests *to the books of the government*, and because the economy is continuing its collapse, this debt is absolutely unpayable.

Let me say it again. There is no way that this debt can ever be paid, so the effect of the bailout will be to bankrupt governments. The real purpose of the bailout is to finish the destruction of the governments that the Inter-Alpha process began decades ago.

In the second decade of the 21st century in Europe you have seen, for example, all of the negotiations over trying to save *debt*, to renegoiate debt, and one nation after another are falling to corporatist dictatorship. Control is the real reason for the bailout. You are watching the nations of the world being destroyed and being replaced by fascist (corporate)dictatorships – an international global fascist imperial dictatorship.

In Europe, the bank bailout followed this pattern. European banks were among the main beneficiaries of an around $16 trillion bailout that came from various Fed funds in 2008. For instance, six of the first eleven beneficiaries of the Term Auction Facility (overnight Fed funds) were European banks, including Société Générale and Royal Bank of Scotland. This unseemly operation was undereported and no one asked any hard questions. How did the Federal Reserve justify this?

It appears that the Federal Reserve had their reasons, because they simply lied about what they were actually doing. The public was told one thing, but not what was actually being done. The Fed was clearly trying to save a system – the corporate imperial system. They are not trying to save just US banks or they wouldn't give all this money to foreign banks. They worked to save the *system*, which is what they feel they have to do.

With the global system swamped with derivatives, if it blows up anywhere, it blows up everywhere. So, they have to protect the weakest links, it's like putting out house fires. You have to put the fire out, wherever it breaks out, otherwise the whole place can burn down.

Please understand: the billion dollar bailouts were never meant to protect the economy, be it of the United States or of Europe. The bailouts have always been about protecting a monopolistic financial imperium and institutionalizing (obtaining government "guarantees") the derivatives casinos. It is about promoting the disease and sacrificing the patient.

* * *

Another thing I am personally witnessing in 2013 in Europe, especially in Spain, but not only limited to Spain is the fact that the banks are buying a great deal of sovereign debt. The question many would ask is: Can that debt ever be paid? If not, why are they doing this?

If you and I wish to borrow money from a bank, the first thing the bank would want to know is, *can we pay it back*? However, in a parallel world, the world of World Company Incorporated, it appears the banks are buying debt because they wish to bankrupt nations. It is an old Venetian shell game: have your adversary go so far in debt to you that you then control them.

The banks, today, are buying debt that is worthless. But, then maybe, many of these banks are not intended to survive.

If you take the oligarchy at its word, the Empire says they wish to reduce the population down to 1-2 billion people on this Planet. Which means that huge portion of the banking system, huge part of national economies, and huge amounts of the population are "destined" to disappear.

What is absolutely certain is that we are currently in a process of complete economic collapse. Is this disintegration accidental, or the result of poor planning, or did it happen on purpose? And if so, by whom and why?

It's both. I say there are different groups with different interests. French economist Jacques Attali, in public interviews, said that the euro was conceived as a deliberate mechanism to impose political union – with which no one agreed. In other words, there may have been a gross error made by creating the euro during a crisis to impose a political dictatorship. The same way that we see today with the European Stability Mechanism.

One intention has been the de-industrialization of Europe and if you look at the European Commission's policy regarding Spain, for example, you see that they canceled all programs that have allowed any Spanish recovery, such as cutting 25% of the science budget. I think it's very clear that nobody in the EU intends for Spain to recover and, in fact, I think a real goal is the reduction of population and a return to a type of feudal structure.

Please understand, the past never comes back. This is what the bailout really is about. In trying to save debt, which is in reality, virtual money (it does not exist), the system may be manipulated so that the debt can never be paid. It all becomes a sham using inflated Monopoly money.

By trying to save "the system," we can *destroy ourselves, destroy the social order* and *destroy the nations*. And that's what the bailout really represents. That's why you see the elite promoting it, because many of these banks will be thrown away anyway. *But the nation-states need to go first*. Once the nations are gone, then the imperium can reorganize itself, however it wants, and set up a new monetary system.

Is there a solution? Of course there is. The solution is to give our economies a purpose, and for each nation on this planet to engender in every person a sense of participation in this great common interest of mankind: to give us all a stake in the sustainability of our global system.

And, we simply need to take this derivatives garbage, which is credit upon credit, upon credit – and cancel it. Wipe it out. Eliminate it. *Just like that!* Derivatives are gambling instruments. And gambling debts, when lost, can be canceled.

Therefore, we do not need to pay this gambling debt, whether they are "incentives" or "financial derivatives." The current system is bankrupt, both morally and financially, and to save the world, we need leaders who are prepared to stand up and put the entire financial system into receivership.

There is something else you need to understand about the way money works in order to understand the current financial crisis.

Credit System vs. Monetary System

The world today is run by monetary systems, *not by national credit systems*. You do not want a monetary system to run the world. You want sovereign nation-states to have their own credit systems, which is a system of their own currency. Above all, you want productive, non-inflationary credit creation by the state, as is firmly stated in the US Constitution. This sound fisical policy of credit creation by nation-states has currently been excluded by Maastrich Treaty from even being considered as economic and financial stratagem for Europe.

A *monetary system* is a creation of the financial oligarchy that basically treats humanity like cattle. It is how the elite have operated for centuries. Oligarchies exist by controlling the "coin of the realm," they control its price and its availability, thereby controlling the people. They use their money control to manipulate our world.

This is the system of Empire, the system that Alexander Hamilton challenged when the United States was created and the U.S. credit system invented. Hamilton said,

> We are not going to ask you for money, we are a sovereign country, we create our own money. We will create credit and we will work in the economy in order to increase the productivity of our people. We will fund infrastructure

projects, manufacturing projects, will fund things that increase labor productivity and make the economy more productive and therefore richer.

This is how real wealth is created: the production process. Instead of borrowing money from the oligarchs, you create your own money as a sovereign nation and use it to escape from the clutches of the oligarchy. And this is what allowed the United States to resist the continuous attempts of the British Empire to recover the US as imperial protectorate. The difference between the monetary and credit system is the principle on which the current oligarchic system is based.

Today, in 2013, in Europe, this non-inflationary credit cannot be produced, because in Europe the governments are subject to control by private banking interests, called independent banking systems. These institutions have the power to regulate government, and to dictate terms to government.

Think about how the institution within this European edifice called European Central Bank functions. It operates like a European independent central bank, which has no government, because there is no government. There is no nation. It's a group of nations run by a private bank.

Don't you see? It is insanity. The supposed "independence" of the Central Bank is a decisive control mechanism for private financial interests, which in Europe have historically been installed as an authoritative instrument against an economic policy of sovereign governments oriented towards the General Welfare of its populations. European banking is a remnant of a feudal society, in which private interests – as typified by the ancient Venetian cartels that went into the shadows in the 14th century.

The End

The worldwide fight we are witnessing today is not for the survival of central banks or the euro, but it is a fundamental fight between sovereign governments and the oligarchic financial system that benefits a small elite. Any nation which cannot control its own currency is not sovereign, and any nation which is not sovereign is vulnerable to assault and subversion by this oligarchy.

Now, if people are to participate in self-government, they must participate in the ideas by which society is self-governed. This would mean the end of oligarchies. Nations that foster the creative-mental development of their population produce a people that will not tolerate oligarchic forms of rule indefinitely. An illiterate, technologically backward populations will. In fact, there is little doubt that illiteracy and technological backwardness has contributed to the emergence of oligarchic rule.

The ideas of a nation-state republic and progress are joined at the hip.

Like many ideas of human creativity, the nation state is not a undertaking to be taken lightly, nor is it designed for immediate desires. It is a long-standing human endeavour that has been designed to extend our sense of self far beyond the confines of our perceptions and feelings of personal well-being. The nation-states connect us to all the generations, before and after, which strive for the human legacy of freedom, liberty and pursuit of happiness.

Endnotes

1   Matt Rodell, Aug. 20, 2009 *Nature* magazine.

2   http://www.larouchepub.com/eiw/public/2012/eir-v39n17-20120427/53_3917.pdf.

3   "The Policy of Controlled Disintegration, Richard Freeman," *EIR*, October 15, 1999.

4   Daniel Yergin and Joseph Stanislaw, "The Commanding Heights: The Battle for the World Economy," *Free Press*, 1997, pp.60-64.

5   Will Banyon, "Rockefeller Internationalism," *Nexus*, Volume 11, Number 1, December-January 2004.

6   Jeremiah Novak, *Atlantic Monthly*, July 1977.

7   "The policy of controlled disintegration," Richard Freeman, *EIR* Volume 26, number 41, October 15, 1999.

8   Ibid.

9   "A History of Rigged & Fraudulent Oil Prices (and What It Can Teach Us about Gold & Silver," Mr. Lars Schall interview with F. William Engdahl, chaostheoren.de.

10   Ibid.

11   John Hoefle, British Geopolitics and the Dollar, *EIR*, May 16, 2008, pp 51-52.

12   John Hoefle, "The End of the Line for the Anglo-Dutch System," *EIR*, March 28, 2008.

13   Ibid.

14   Pierre Beaudry, "Mennevee Document on the Synarchy Movement of Empires," Book IV.

15   John Hoefle, "Southern Strategy, Inc: Where Wall Street Meets Tobacco Road," *American Almanac*, February 2001.

16   Lyndon LaRouche, "It's the British Empire, Stupid!," *EIR*, January 11, 2008.

Chapter Two

Genetically Modified Conspiracy

As I write these lines, over a billion people around the world are going hungry and starving. This is only the beginning. There are roughly two billion people across the world that spend more than 50 percent of their income on food. The effects of the 2007 economic meltdown has been staggering: 250 million people joined the ranks of the hungry in 2008 – a number never seen before.

> In a sense it's a genuine paradox. Our planet has everything we need to produce nutritious natural food to feed the entire world population many times over. This is the case, despite the ravages of industrialized agriculture over the past half-century or more. Then, how can it be that our world faces, according to some predictions, the prospect of a decade or more of famine on a global scale? The answer lies in the forces and interest groups that have decided to artificially create a scarcity of nutritious food.[1]

World Trade Organization

One of the organizations most responsible for this tremendous growth in world hunger is the World Trade Organization. Created in 1994 out of the GATT Uruguay Round, the WTO introduced a radical new international agreement, Trade-Related Aspects of Intellectual Property Rights" (TRIPS), which permitted multinationals to patent plant and other life forms for the first time.

The WTO was created after World War II by internationalist in Washington to serve "as a wedge to push free trade among major industrial nations, especially the European Community."² It was given birth out of wedlock on January 1, 1995 when the Marrakech Agreement replaced GATT, which had commenced in 1948.

According to their own publicity, the World Trade Organization establishes a framework for creating non-discriminating, reciprocal trade policies. The reality appears quite different. The WTO's anti-nation-state intent can be readily seen in their 1988 slogan: "One World, One Market." That slogan came from the GATT Montreal summit of its predecessor, Uruguay Round of Agriculture "Reform" (1986-94) – the process in which WTO was birthed.

A watershed moment came in 1993 when the European Union agreed to a major reduction of their national agriculture protection.

This reduction was a multi-stage process. First step, according to the incoming WTO game rules, the members nations were to be forced to open their borders, to grant the right to operate freely within those borders to other nations, and to eliminate national grain reserves.

Grain reserves were no longer to be the dominion of independent nation-states. They became property, to be managed by the "free market," by private, mostly American mega-corporations, in other words, corporations running the world markets.

These companies were already dominant, but now, "they had a new unelected supranational body to advance its private agenda on a global scale. WTO became a policeman for global free trade and a (predatory) battering ram"³ with an annual budget worth trillions of dollar. "Its rules are written with teeth for punitive leverage to levy heavy financial and other penalties on rule violators."⁴ Under this regimen, agriculture control is a priority.

What's more, the rules, which were sold as the beacon of hope for underdeveloped countries, were written by the corporate giants that form the nucleus of World Company Inc. The blueprint for "market-oriented" agricultural reform was written by D. Gale Johnson of the University of Chicago for David Rockefeller's Trilateral Commission, and former Cargill executive Dan Amstutz played a prominent role in drafting the agriculture rules of the GATT Uruguay Round. Cargill is the world's largest grain company.

New draconian rules have been forced upon a free and integrated global market to control its products. The new agreements have also banned agricultural export controls, even in times of famine. Today, the cartels domination over the world export grain trade has grown even greater.

Further, this international pact forbids countries from restricting trade through food safety laws called trade barriers. This stratagem has also opened the world markets to unrestricted GMO food imports with no need to prove their safety, but more on that, later.

Agriculture is food, and food is what we eat. We too often take it for granted – especially in the first world – that we will always have an abundance of food. A trip to the supermarket, and our needs are taken care of. What happens if one day soon, you wake up, and there is nothing to eat? Then what?

WTO propaganda tells us that the world "market" and so-called "free-trade" will somehow provide the favorable conditions necessary for growth. However, with the creation of multi-nation trading zones, citizen-responsive governments *lose power and control* at the expense of these supra-national agencies that supersede the authority of nation-states. These agencies do not in any way represent the people of any one nation. Instead, their loyalties lie with the corporations and financial organisms that elect them, fund them and support them. These organizations form one of the nodes of elite rule, a top down imperious structure that seeks to enslave populations through many guises, including subtle pschyological warfare and other soft-war techniques.

Agreements such as GATT and NAFTA have subtly destroyed national economies by putting them under the imperatives of world commerce and globalization without borders. Globalization is a top down concept, which means that the farmers – the people, who actually put the food on our table – are wiped out and replaced by giant multi-faceted corporations, which strive to control the production and distribution of food.

During the last two decades, millions of farmers in the United States, Europe, Canada, Australia, and Argentina have been wiped out. For example, in 1982, the United States still had 600,000 independent hog farmers. Today, that number is less than 225,000.⁵

Make sure you understand – this is no accident – fewer independent farmers mean greater corporate control over what you eat. People can pretty much get used to anything in life, except not having enough to eat. Even death is easier. You only have to suffer it once.

GATT, North American Free Trade Agreement (NAFTA), Central America Free Trade Agreement (CAFTA) and every such binding agreement has helped spawn ghettos and shanty towns in cities throughout Latin America, Asia and Africa by creating conditions that force people off their land, while elite take over the means of production. Shanty towns and ghost towns equal depopulation on the one hand. On the other, if you can force people off their land and into the cities, you are creating a perfect storm of discontent amongst the masses. Mass unrest requires armed forces control – problem, reaction, solution.

This is exactly the prediction of UK's Ministry of Defense *Strategic Trends* report, which, as we have said, was based on a 2005 Bilderberg blueprint. In 2013, over 50% of the world's population is living in urban rather than rural environments. The report states:

> There will be a substantial growth in shanty towns and unplanned, random urban settlement, increasing the resource cost and environmental impact.⁶

With the destruction of nation-state republics and the creation of mega-economic blocks, which are linked to each other through a globalized marketplace, independent countries can be "replaced by Mega-Cities" with population bases of over 20 million people.

Caused by a massive population displacement, these cities will swell to unimaginable proportions "which will already have experienced endemic lawlessness and high levels of violence."

One of the first experiments to bring about the depopulation of large cities was conducted in Cambodia by the Pol Pot regime. It is interesting to note that a model for Pol Pot's genocidal plan was drawn up in the United States by one of the Club of Rome's supported research foundations, and overseen by Thomas Enders, a high-ranking State Department official.

The Club of Rome, founded in 1968, is made up of some of the oldest members of Venetian Black Nobility. The Club is the most important institution in the world to push the Malthusian depopulation scheme. A report by the Club of Rome leaves little room for doubt as to their real agenda:

> In search for a new enemy to unite us, we came up with the idea that pollution, the threat of global warming, water shortages, famine and the like would fit the bill.

They concluded with the following:

> The real enemy, then, is humanity itself.

Thus, a leading international institutions is pushing policies of retrogression in technology, and reduction of the world's population by several billion people – that's genocide, in case you didn't know it.

And there is no better or cheaper way to reduce population than through starvation. And in order to starve a people to death, you must take control of their food production away from independent farmers, and put it into the hands of giant corporations subservient to the interests of World Company Inc. – some refreshing food for thought the next time you have your breakfast cereal.

Within the WTO regulations, nations are prohibited from protecting local economy or tax goods, even when these goods clearly were produced with slave labor. Furthermore, nations are not allowed to give preference to local economies that hire local people at decent wages to produce goods that then benefited the local business and economy – people who paid taxes, played by the rules and reinvested their hard earned wealth into the local or national market.

The truth is: "free trade is rigged trade, and the 'fairness' question is diversionary propaganda for deluded lawmakers, farmers, and the public. It is run by an international financial cartel. The cartel interests control the playing field: who plays and the rules of the game."[7]

When you expand the cartel's control into strategic areas such as food, the situation gets serious in a hurry.

Food as Weapon

The result of the imposition of WTO rules, internationally, over the past 15 years has been that food processing and trade have come to be monopolized to an extreme degree by a small tight clique of cartel companies. These firms dominate international commodity flows, and even the domestic food supplies and distribution of most nations.

The use of food as a weapon is an ancient practice and can be found at least four thousand years ago in Mesopotamia. In ancient Greece, the cults of Apollo, Demeter and others often controlled the shipment of grain and other foodstuffs through a temple system.

Imperial Rome, Venice, the powerful Burgundian duchy, the Dutch and British Levant companies, the East India companies, and West India companies – all followed suit. Today, food warfare is firmly under the control of just a few corporations.

The largest food company in the world is Nestlé. It was founded in 1867, and is based in Switzerland. It is the number-one world trader in whole milk powder and condensed milk; the number-one seller of chocolate, confectionery products, and mineral water (it owns Perrier); and the number-three U.S.-based coffee firm. It also owns 26% of the world's biggest cosmetic company, L'Oreal.

Much of the rest of the milk powder business is controlled by the Anglo-Dutch-owned Unilever, the result of a 1930 merger of a British and a Dutch firm. Unilever is the number-one world producer of ice cream and margarine, as well as a key player in fats and edible oils. Unilever owns vast plantations and runs Africa's largest trading company, United Africa Co. In Zimbabwe, Congo-Zaire, Mali, Chad and Sudan, the United Africa Co. is projected by the British intelligence as ushering in a "United States of Africa."

"The boundaries among states are to be dissolved, and their contents organized as a new business franchise with two purposes: first, the security of foreign investment and seizure of property titles on raw materials by primarily British Commonwealth mining and other companies;"⁸ and second, the lining of the pockets of the government enforcers of the policy.

These policy enforcers already primarily agree with the ideology of zero growth – the idyllic primitive life of "useless eaters," as Henry Kissinger has called anyone living below the Tropic of Cancer. "Not only in Africa, but Third World peasant populations will be organized to do what has come 'naturally' for years -- apply their quaint picks and hoes to land wasted by centuries of labor intensive farming to scrape together tribute to the World Bank in the form of food. The result will be a net decline in food production and consumption worldwide."⁹

But, it gets worse, much worse.

Minnesota-based Cargill Company is the world's largest grain company. Grain constitutes a dominant portion of the standard diet. Since the 1920s, the billionaire MacMillan family have run Cargill. The MacMillans are members of the Bilderberg Group.

John Hugh MacMillan, president and chairman of Cargill from 1936 through 1960, held the title of hereditary Knight Commander of Justice in the Sovereign Order of St. John (Knights of Malta), one of the Vatican's most important orders.

> Cargill's international trading arm, Tradax, Inc., is headquartered in Geneva, Switzerland. The Lombard, Odier Bank, as well as the Pictet Bank, old, private and very dirty Swiss banks, owns a chunk of Tradax. The principal financier for Tradax is the Geneva-based Crédit Suisse, which is one of the world's largest money-launderers.
>
> Archer Daniels Midland's purchase of Töpfer, a Hamburg, Germany-based grain company vastly increased ADM's presence in the world grain trade.¹⁰

Then, there are the seed companies. By far the biggest and the dirtiest of them all is Monsanto, with an international workforce of 21,035, in 404 facilities in 66 countries. Monsanto has power that supersedes the influence of most governments on the planet. An associate of Monsanto, Dr. Roger Beachey, was appointed by President Brack Obama in 2010, as the Science Advisor to the Agriculture Department. This is a good example of the interlocked financial, economic, political and business interests dominating the food industry.

The DuPont Chemical Co., owns Pioneer Hi-Bred International Inc., an Iowa based largest seed-corn company in the world. It sells a range of crop and forage seeds in 70 countries.

Syngenta is based in Switzerland and operates in 90 countries, with a workforce of 26,000. It was formed in 2000 from the merger of Novartis Agribusiness and Zeneca Agrochemicals. Novartis was itself formed by the merger of the legendary Swiss chemical firms, Sandoz and Ciba-Geigy. Zeneca Agro came out of the British firms ICI (Imperial Chemical), and AstraZeneca.

DowAgroSciences LLC is based in Indianapolis and is a subsidiary of The Dow Chemical Co. It was formed in 1997 as a joint venture between Dow's agriculture sciences division and the Eli Lilly Co.

BASF Plant Science, based in Germany, was established in 1998, as a centralization of all the agriculture bio-technology capacity of the longstanding BASF Chemical Co. BASF Plant Science has a 700-person research effort, focusing on plant genetics and patentable traits, in collaboration with the mega-seed companies.

Bayer CropScience is also based in Germany, and is the second-largest pesticide firm in the world. It operates in 120 countries, with 20,700 employees, making and selling fungicides, insecticides, and other plant protections, while also working on new bio-engineered formulate.[11]

In other words, ten to twelve pivotal companies, assisted by another three dozen, run the world's food supply. They are the key components of the Anglo-Dutch-Swiss food cartel, which is grouped around Britain's House of Windsor. Led by the six leading grain companies – Cargill, Continental, Louis Dreyfus, Bunge and Born, André, and Archer Daniels Midland/Töpfer – the Windsor-led food and raw materials cartel has complete domination over world cereals and grains supplies, from wheat to corn and oats, from barley to sorghum and rye. But it also controls meat, dairy, edible oils and fats, fruits and vegetables, sugar, and all forms of spices.

The first five of the companies are privately owned and run by billionaire families. They issue no public stock, nor annual report. They are more secretive than any oil company, bank, or government intelligence service.

While these firms maintain the legal fiction of being different corporate organizations, in reality this is one interlocking syndicate, with a common purpose and multiple overlapping boards of directors. The Windsor-centered oligarchy owns these cartels, and they are the instruments of power of the oligarchy, accumulated over centuries, for breaking nations' sovereignty. To understand the reality as opposed to the rhetoric of their involvement in world economy, it is better to study what companies do rather than what they say.

Cartel's Big Six grain trading companies own and control 95% of America's wheat exports, 95% of its corn exports, 90% of its oats exports, and 80% of its sorghum exports. The grain companies' control over the American grain market is absolute.

The Big Six grain companies also control 60-70% of France's grain exports. France is the biggest grain exporter in Europe (the world's second largest grain exporting region), exporting more grain than the next three largest European grain-exporting nations combined.

In sum, the Anglo-Dutch-Swiss food cartel dominates 80-90% of the world grain trade. In fact, however, the control is far greater than the sum of its parts: The Big Six grain companies are organized as a cartel; they move grain back and forth from any one of the major, or minor, exporting nations. Cargill, Continental, Louis Dreyfus et al. own world shipping fleets, and have long-established sales relationships, financial markets, and commodity trading exchanges (such as the London-based Baltic Mercantile and Shipping Exchange) on which grain is traded, which completes their domination. No other forces in the world, including governments, are as well organized as the cartel.[12]

Monsanto, Cargill, Archer Daniels Midland (ADM) became Household Names

By 2004, the four largest beef packers controlled 84% of steer and heifer slaughter – Tyson, Cargill, Swift and National Beef Packing; four giants controlled 64% of hog production – Smithfield Foods, Tyson, Swift and Hormel; three companies controlled 71% of soybean crushing – Cargill, ADM and Bunge; three giants controlled 63% of all flour milling, and five companies controlled 90% of global grain trade; four other companies controlled 89% of the breakfast cereal market – Kellogg, General Mills, Kraft Foods and Quaker Oats; in 1998, Cargill acquired Continental Grain to control 40% of national grain elevator capacity; four large agro-chemical/seed giants controlled over 75% of the nation's seed corn sales and 60% of it for soybeans while also having the largest share of the agricultural chemical market – Monsanto, Novartis, Dow Chemical and DuPont; six companies controlled three-fourths of the global pesticides market; Monsanto and DuPont controlled 60% of the US corn and soybean seed market – all of its patented GMO seeds; and 10 large food retailers controlled $649 billion in global sales in 2002, and the top 30 food retailers account for one-third of global grocery sales.[13]

Please understand, this interlocked self-perpetuating syndicate decides who eats and who doesn't, *who lives and who dies*. It is a virtual spider web of financial, political, economic and industry interests with the Venetian ultramontane fondi model at the center. These people own and manage the affairs of an interlocking corporate apparatus that dominates choke points within the global economy, especially finance, insurance, raw materials, transportation, and consumer goods.

Cargill, the largest privately held agribusiness corporation, along with Archer Daniels Midland have became the arbiters of death. The question is: Why are mega corporations and a small socio-political elite allowed to own our food – to control the very basis of human survival?

How it Works

The control works as follows: The oligarchy has developed four regions to be the principal exporters of almost every type of food, in the process acquiring top-down control over the food chain in these regions. These four regions are: the United States; the European Union, particularly France and Germany; the British Commonwealth nations of Australia, Canada, the Republic of South Africa, and New Zealand; and Argentina and Brazil in Ibero-America. These four regions have a population of over a billion people, or 15% of the world's population. The rest of the world, with 85% of the population – 4.7 billion people – is dependent on the food exports from those regions.[14]

Can the nations protect themselves? Not if they are members of World Trade Organization. If any country tries to protect its local markets, then the entire world community is entitled to rebel against the "protectionist policies."

As economist William Engdahl writes in *The Seeds of Destruction*:

> The WTO rules assert that nations must eliminate food reserves, eliminate tariffs on food imports and exports, cease intervening to support their domestic farm sector – all under the imperial rationalization that such nation-serving measures would be 'trade-distorting' practices, which would impede the free-market 'rights' of the globalist corporations. Now one-seventh of the world's population lacks enough to eat. Against this backdrop, the story of the WTO is one of crimes against humanity, and not an academic economics debate."
>
> In the face of mass death, through starvation, through lack of food, recall what the core WTO liturgy is: Nations must not keep food reserves, because this would be trade-distorting. Nations must not attempt to be food self-sufficient, because this would deny their citizens the right to access the world market. Nations must not support their own farmers, because this harms farmers elsewhere. Nations must not use tariffs, because this denies right-of-access to your citizens by foreign producers. And on and on and on. The consequences of these actions are genocidal so don't debate it. Cancel it.[15]

There is also another area being utilzed by the elite to take control of our independence – devaluation of the currency. And the devaluation of currency is directly related to our purchasing power. This became part of the scheme under Nixon's New Economic Plan (NEP) that included closing the gold window in 1971 and the invalidation of the Bretton Woods agreement.

Under Bretton Woods, at the close of World War II, a gold reserve standard was established, with the U.S. dollar pegged to gold at $35 an ounce. Bretton Woods completely eliminated the risk of dramatic currency losses as a result of speculative runs on national currencies.

Once the federal gold window was closed and the Bretton Woods agreement scrapped, currencies were allowed to float freely. Developing nations became targets, because the corporate elite and the World Company Incorporated could not allow these nations to attain food-sufficiency in grains and beef.

Instead, by forcing them to rely on America for key commodities, and with the dollar value manipulated at will, Third World nations were forced to concentrate on small fruits, sugar and vegetables for export. Then with that earned foreign exchange they could then *purchase* some of their needed goods plus *pay* IMF and World Bank loans for the rest, creating a never-ending cycle of debt slavery.

Today, nothing has changed, except that the currency debasement used against the Third World forty years ago is now being actively and openly used against the helpless populations in America and Europe. For example, a small county of 100,000 people in the United States between the years 2003 and 2008 lost over $3.3 billion of purchasing power. Again, please understand, to debase a currency, is to reduce its value and purchasing power.

Money does effect peoples' lives. The United Nations has estimated that the 2007 banking crisis drove 100 million people back into poverty around the world. The mortality statistics and morbidity statistics – the number of extra people who die and are ill – rise dramatically when a population lives in poverty. Paul Moore, former head of risk for the Halifax Bank of Scotland stated:

> He would be very surprised if the financial crisis didn't kill more people than any single conflict since WWII.[16]

Please never forget that any government that allows its internal market to turn into a free-trade market for the world in food, in the name of the deceptive "economic efficiency," is consciously deterimined to kill its own population.

This policy serves to directly undercut the purpose of a government to promote the general welfare of it's citizens, and it furthers the intent of the corporate imperial system, to drastically reduce world agro-industrial potential and create conditions for depopulation. Under the "markets" principle, the "global sourcing" of food by corporate monopolies, works to the detriment of the populations in both the exporting and importing nations.

It may seem difficult to understand, but it is easy to explain.

As I said in the first chapter, there are currently seven billion people on Planet Earth, a small planet with limited natural resources and an ever expanding population base. Food and water are becoming ever scarcer. *For the elite to eat, you and I have to die.*

Needless to say, the policies and programs that push the depopulation issues are worked out at a supranational level. Behind the scenes and away from public spotlight, private and very secretive organizations such as the Trilateral Commission, Council on Foreign Relations and the Bilderberg Group take care of business, making sure nation-states abide the free-trade line.

Trilateral Commission

Without a doubt, the Bilderberg Group is the premier secret forum operating in the shadows of power, but the Trilateral Commission also plays a vital role in the One World Incorporated's scheme to use wealth, concentrated in the hands of a few, to exert world control. And as I have said before, there is no better way to control an ever growing population than through the only resource people cannot do without – food.

The Trilateral Commission, with easily recognizable members, was founded in 1973 at a meeting attended by 300 influential, handpicked Rockefeller friends from North America, Europe and Japan.

Zbigniew Brzezinski, Jimmy Carter, Gerald Ford, George H.W. Bush, Paul Volcker (Carter's Federal Reserve Chairman) and Alan Greenspan answered David Rockefeller's call for creating a:

> … community of a developed nations, which can effectively address itself to the larger concerns confronting mankind.[17]

The organization "laid the basis for a new global strategy for a network of interlinked international elites," many of whom were Rockefeller business partners. Combined, their financial, economic and political clout was unmatched. So was their ambition.

From William Engdahl's *Seeds of Destruction*:

> Trilateralists laid the foundation for today's globalization. They also followed Samuel Huntington's advice that democracy's unreliability had to be checked by some measure of (public) apathy and non-involvement (combined with) secrecy and deception.
>
> The Commission further advocated privatizing public enterprises along with deregulating industry. Trilateralist Jimmy Carter embraced the dogma enthusiastically as President. He began the process that Ronald Reagan continued in the 1980s almost without noticing its originator or placing blame where it's due.

Slyly and without much noise, a movement towards a larger control mechanism, through a variety of indirect pacts that were already developing limitations on national sovereignty, supranational control was being cobbled together, one agreement at a time.

In the aftermath of World War II, world agriculture domination, along with controlling world oil markets, was to be a central pillar of Washington foreign policy. Another pillar had a name, it was called the Green Revolution.

Post WWII

Franklin Delano Roosevelt died. Harry Truman became President. The War ended. Japan capitulated under the weight of the horrors of an atomic explosion. The Soviet Union became a superpower. Winston Churchill made his famous speech. The

Iron Curtain went up, and the world became bipolar. The Cold War was upon us. Food became a weapon – through stealth.

During the Cold War, food was a strategic weapon. Masquerading as "Food for Peace," it became a cover for US agriculture to engineer the transformation of family farming into global agribusiness. A goal being the elimination of the small farmer, in favor of the mega-corporations

> The shortage of grain staples along with the first of two 1970s oil shocks advanced a significant new Washington policy turn. The defining 1973 event was a world food crisis. Oil and grains were rising three to fourfold in price when the US was the world's largest food surplus producer with the most power over prices and supply. It was an ideal time for a new alliance between US-based grain trading companies and the government. It 'laid the groundwork for the later gene revolution.'[18]

But before gene revolution, we had green revolution, with the usual panache and pomp announcing to the world that starvation eradication was just around the corner. Nobody at the time realized that corners are never round, but that's beside the point.

GREEN REVOLUTION

The Green Revolution, in a publicity campaign touted by the mainstream media, was purported to have saved over a billion people from starvation in Mexico, Latin America, India and other select countries, and with its free market efficiency during the 1950s and the 1960s, the "revolution" increased agriculture production around the world. According to the propaganda:

> ... it involved the development of high-yielding varieties of cereal, expansion of irrigation infrastructure, modernization of management techniques, distribution of hybridized seeds, synthetic fertilizers, and pesticides to farmers.[19]

In reality, as it emerged years later, the Green Revolution was a brilliant Rockefeller family scheme to develop a globalized agribusiness, which they then could monopolize just as they had done in the world's oil industry beginning a half century before. It was called agribusiness, in order to differentiate it from traditional farmer-based agriculture – the cultivation of crops for human sustenance and nutrition.

It gave US chemical giants and major grain traders new markets for their products. Agribusiness was going global, and Rockefeller interests were in the vanguard helping industry globalization take shape.[20]

Economist William Engdahl explains:

> A crucial aspect driving the interest of the Rockefeller Foundation and US agribusiness companies was the fact that the Green Revolution was based on proliferation of new hybrid seeds in developing markets. One vital aspect of hybrid seeds was their lack of reproductive capacity. Hybrids had a built in protection against multiplication. Unlike normal open pollinated species whose seed gave yields similar to its parents, the yield of the seed borne by hybrid plants was significantly lower than that of the first generation. A handful of company giants held patents on them and used them to lay the groundwork for the later GMO revolution. Their scheme was soon evident. Traditional farming had to give way to High Yield Varieties (HYV) of hybrid wheat, corn and rice with major chemical inputs.
>
> It was the beginning of agribusiness, and it went hand-in-hand with the Green Revolution strategy that would later embrace plant genetic alterations. The 'Revolution' also harmed the land. Monoculture displaces diversity, soil fertility and crop yields decrease over time, and indiscriminate use of chemical pesticides causes serious later health problems. That began the process of debt enslavement from IMF, World Bank and private bank loans. Large landowners can afford the latter. Small farmers can't and often, as a result, are bankrupted. That, of course, is the whole idea.
>
> One key effect of the Green Revolution was to depopulate the countryside of peasants who were forced to flee into shantytown slums around the cities in desperate search for work. That was no accident; it was part of the plan to create cheap labor pools for forthcoming US multinational manufactures, the 'globalization' of recent years.

> The Green Revolution was typically accompanied by large irrigation projects which often included World Bank loans to construct huge new dams, and flood previously settled areas and fertile farmland in the process. Also, super-wheat produced greater yields by saturating the soil with huge amounts of fertilizer per acre, the fertilizer being the product of nitrates and petroleum, commodities controlled by the Rockefeller-dominated Seven Sisters major oil companies.
>
> Huge quantities of herbicides and pesticides were also used, creating additional markets for the oil and chemical giants. As one analyst put it, in effect, the Green Revolution was merely a chemical revolution. At no point could developing nations pay for the huge amounts of chemical fertilizers and pesticides. They would get the credit courtesy of the World Bank and special loans by Chase Bank and other large New York banks, backed by US Government guarantees.[21]

Note and please remember, the chemical fertilizers and pesticides are in the hands of the same cartels, which control access to food. These same cartels claim patent rights to seeds and crop traits. The same cartels also control access to the technologies that manipulates the genetic characteristics, and with full Washington and WTO backing, they are playing God and patenting life. These are the same cartels that also control the production and distribution of what we eat, as well as the shipping routes that deliver the food to us.

But it does not end there.

The control of food supplies is a matter of national security. The U.S. Department of Agriculture is one of the key elements in a national security edifice attempting to control the world food market.

Genetically Modified Organisms

On a personal level, I first became aware of the elite's focus on agriculture after I began looking into genetically modified crops. I soon became interested in *how* the global agricultural markets worked. The more I researched, the more I realized that one of the elite's best weapons being promoted in the corporate media was genetically modified seeds – seeds with a genetically engineered

DNA and modified through the introduction of foreign bacteria resistant genes, creating new species of seed corn, soybean, and so on.

This was being presented as *the* solution to world hunger, but to me, none of it made much sense. I wondered where the whole idea of genetically modified crop came from. And lo and behold, I came across a well-known name, an entity that was known for their efforts to control oil and global power: the Rockefeller family and the Rockefeller Foundation.

In fact, there are three or four giant agribusiness and agrochemical groupings, rooted in the chemical industry, which have virtually monopolized the market and have established a cartel around genetically modified seeds.

And I soon also realized that it was a political issue. The more I delved the more I investigated, the more diabolical the plans for genetic manipulation seemed to me. In fact, what I discovered was that these plans were directly related to the involvement of the Rockefellers in what is called eugenics, the racist policy also used by Hitler and the Nazis in the Third Reich.

But, what few people talk about, is that many of the Nazi's eugenics' programs were funded by the Rockefeller Foundation. The affiliation between genetics and eugenics has remained intact from then until now, with on-going corporate involvement.

Monsanto is one of these companies, and like Haliburton or Exxon Mobile or Boeing, which are considered national security assets, they take advantage of the resources of the U.S. state and international institutions to increase their power over the global economy.

The Rockefellers funded the early research that created the transgenic products through the introduction of foreign bacteria into specific strains of corn or soybeans. But the only strain marketed, so far, on a large scale has been the seed resistant to Monsanto's Roundup. Roundup, Monsanto's glyphosate-based herbicide, has become the world's best selling weed-killer.

Monsanto's GM seeds are the only seed strains that resist the toxins that are sprayed on crop and kill everything around them. The corn stalks resistant to toxins give the impression of strength and vitality. All of this is a part of a Rockefellers' wet dream –

they start with seeds and plants, but ultimate they manipulate human genes.

One of their plans deals with financing a project to develop a type of maize, a key staple in Latin America, which contains an element that causes the human sperm not to be able to conceive children. If this is not a eugenic plan for population reduction, I do not know what is. The project is supported and funded by the U.S. government through the Department of Agriculture.

Henry Kissinger

In the early 1970s, Nixon, busy with the Watergate affair, had little time for the business of the presidency. The acting Commander-in-Chief, according to some, was Henry Kissinger, who in April 1974 issued National Security Study Memorandum 200. This was a top-secret report on national security, which stated that the reduction or control of populations was a prerequisite for the United States to provide food aid and other assistance to foreign countries.

The aim was to adopt a plan for drastic global population control – that is, to reduce it to three billion people by the year 2050.

Genetically Modified Food

GM foods today saturate our diet. In all supermarkets in the developed world, one finds Nestlé products, Monsanto, Unilever and all major brands. All of these companies promote and sell GM foods because it appears that 70% or more of the products that Americans consume contain Monsanto's or other company's GM products.

Everything from potatoes, tomatoes, and corn to rice and wheat; legumes like soybeans; vegetable oils; soft drinks; salad dressings; infant formula; vegetables and fruits; dairy products including eggs; meat and many other animal products that we buy in the supermarket have been genetically modified.

It is a form of control: it ruins local farmers in countries like Argentina and Brazil and replaces them with foreign-owned, giant industrial complexes. Farmers are driven from their land and forced to subsist in the cities, living in slums, constituting cheap labor for global industry manufacturing concerns.

> Food inspection authorities and biologists experimenting with the manipulation of DNA structures for large food companies claim that these products have undergone sufficient testing and form no danger to public health.[22]

However, this is a blatant lie. For example, the world's leading plant lectins and plant genetic modification expert, Arpad Pusztai discovered that:

> ... rats fed GMO potatoes had smaller livers, hearts, testicles and brains, damaged immune systems, and showed structural changes in their white blood cells making them more vulnerable to infection and disease compared to other rats fed non-GMO potatoes. It got worse. Thymus and spleen damage showed up; enlarged tissues, including the pancreas and intestines; and there were cases of liver atrophy as well as significant proliferation of stomach and intestines cells that could be a sign of greater future risk of cancer. Equally alarming – this all happened after ten days of testing, and the changes persisted after 110 days that's the human equivalent of ten years.[23]

Furthermore eating genetically modified corn and consuming trace levels of Monsanto's Roundup chemical fertilizer caused rats to developed an extraordinary number of tumors, widespread organ damage, and premature death, according to research,[24] carried out by Caen University in France,[25] which looked at the long-term effects of consuming Monsanto's genetically modified corn. The research is described as "the most thorough research ever published into the health effects of GM food crops and the herbicide Roundup on rats."

> Despite the enormous risks, however, Washington and growing numbers of governments around the world in parts of Europe, Asia, Latin America and Africa now allow these products to be grown in their soil or imported. They're produced and sold to consumers because agribusiness giants like Monsanto, DuPont, Dow AgriSciences

and Cargill have enormous clout to demand it and a potent partner supporting them – the US government and its agencies, including the Departments of Agriculture and State, FDA, EPA and even the defense establishment.[26]

A Common Denominator

Do you know what Agent Orange, saccharin, bovine growth hormone, GM soybeans and the first nuclear weapons, all have in common? They were created or distributed by Monsanto. Monsanto's first product was saccharin, it was later proved to be a carcinogen. Furthermore, these US GMO giants also have a long sordid association with the Pentagon in supplying massively destructive chemicals like the Agent Orange and napalm used to defoliate Vietnam jungles in the 1960s and 1970s, thereby exposing hundreds of thousands of civilians and US troops to deadly dioxin – one of the most toxic of all known compounds.

Stop and think for a moment. Monsanto, Cargill, DuPont & Co. want us to trust them with the most important things we ingest – *our food*.

Food is power. And when it is used to cull the population, it becomes a weapon of mass destruction. You may not realize it, but the Rockefellers and their friends certainly do.

Today, the entire population is being used as lab rats in an uncontrolled, unregulated mass human experiment for these completely new, untested and potentially hazardous products.

> Food safety and public health issues aren't considered vital if they conflict with profits. And leading the effort to develop them is a company with a long record of fraud, cover-up, bribery, deceit and disdain for the public interest – Monsanto.[27]

But there is an even darker side to the Monsanto GMOs – a depopulation agenda. As in reducing the world's population through what we eat. Surprised?

One of the largest organizations working behind the scenes on depopulation agenda is the Rockefeller Foundation. In 2001, a privately owned biotech company, Epicyte, financed by the

Foundation announced it successfully developed the "ultimate GMO crop" – contraceptive corn. It was called a solution to world over-population.

In a late development, the Food and Drug Administration (FDA) has put into effect new policies that allow genetic engineering companies, predominantly the "Big Agro" corporations, to put on the market any Genetically Modified Organisms they wish without consulting the public. When you think about this, it's not just a GMO grass that takes less water, or GMO tomatoes that last much longer on the shelf. Genetic engineers say that in different studies, GMO potatoes, rice, etc., there could be side effects such as the reducing fertility in second and third generations.

Reduced fertility ... eugenics ... depopulation.

Along with the public traits these organisms are being engineered for, there are other traits quietly being built in, to cause biological changes in the body. In one groundbreaking study conducted by Andres Carrasco, the head of the molecular Embryology Lab at the University of Buenos Aires, and chief scientist at the National Council for Science and Technology, found that pesticides used in crops may be an endocrine disruptor.

> The ingredient glyphosate, which is used in Roundup herbicide, is increasing the number of birth defects in animals. The birth defects range from a condition called cyclopia, in which a single eye develops in the center of the forehead, and other defects like stillbirths, cancer and miscarriages.[28]

Let's back track to the 1980s.

> Rockefeller Foundation funding throughout the 1980s had a particular objective in mind – to learn if GMO plants were commercially feasible and if so spread them everywhere. It was the new eugenics and the culmination of earlier research from the 1930s. It was also based on the idea that human problems can be solved by genetic and chemical manipulations, as the ultimate means of social control and social engineering. Foundation scientists sought ways to do it by reducing infinite life complexities to simple, deterministic and predictive models under their diabolical scheme

- mapping gene structures to correct social and moral problems including crime, poverty, hunger and political instability. With the development of essential genetic engineering techniques in 1973, they were on their way.[29]

In fact, the proliferation of GMOs is the icing on the cake of agro-business. In the fifties, the Harvard business school developed a product called the "agro-industrial project." Ray Goldberg was a professor of the Faculty of Business, and John Davis was advisor to the Department of Agriculture. Davis and Goldberg theorized on a cartel structure for agriculture, based on the system by which the oil majors monopolistically controlled world oil market.

Not surprisingly, the idea came from the Rockefellers. It was a project that moved forward step by step, starting with orange juice produced in Florida. Florida orange producers soon became the weakest link in the chain rather than being the most important. Because with the agro-industrial system, everything was based on efficiency and maximum profitability for the cartel that dominated the top of the pyramid. This system has been perfected through globalization.

Today, there are roughly twenty to thirty mega-business groups, of which four are concerned with GMOs, such as Monsanto and Bayer, the aspirin producer in Germany. But all of them act as a cartel that tries to monopolize all the agro-essential seeds in the world. They will buy independent seed companies, whether they produce good or bad seeds, simply to monopolize the market.

However, the control of the seeds by the agro-industry is but one aspect of the elite's diabolical plan to control us and our world. As is the case with the financial meltdown, orchestrated from behind the scenes by the elite, the record rise in grain and food prices is an integral part of a long-term strategy by the Rockefeller brothers in organizing the global food chain along the same monopoly model they had used in oil and medicine.[30] How was it done?

De-Regulation and Speculation

In 2007, everything came to a screeching halt, the financial bubble exploded and speculators began losing money by the billions.

The Wall Street beast needed to be fed, and speculators and pundits clamored for greater and freer trade to offset the financial meltdown. The Ponzi schemers, the speculators and the special interest groups pushed for fully-deregulated markets. Everything was to become fair game. The world was up for sale. The US government's push to fully deregulate wheat exports betrayed the nation's wheat farmers to the predatory global grain cartel and commodity speculators.

Deregulating grain speculation and creating favorable conditions for unchecked market manipulations by eliminating agricultural commodity derivatives regulation became the game in town – a new game with boundless possibilities.

> The historic and unprecedented deregulation opened a massive hole in Government supervision of derivatives trading, a gaping hole that ultimately facilitated the derivatives games leading to the 2007 US sub-prime financial collapse.
>
> Of course, deregulation by government merely opened the door to private control – another form of regulation – by the largest and most powerful corporate groups in any given industry. That was certainly the case for agriculture – the big four-grain cartel companies dominated world grain markets from the 1970s to today. They worked hand-in-glove with big Wall Street derivative players such as Goldman Sachs and JP Morgan Chase and Citigroup. By the latter part of 2007, trading in food derivatives was fully deregulated by Washington, and US government grain reserves gone. The way was clear for dramatic food price rises.[31]

Back in 1991, Goldman Sachs came up with its own commodity index, a derivative that tracked all commodities from metals to energy and from oil to food, soy and wheat. Wall Street greed took over and with it, the ability to manipulate the price of essential foods worldwide at will.

The Goldman Sachs Commodity Index reduced food to a mathematical formula. Barclays, Deutsche Bank, JP Morgan Chase, AIG, Bear Stearns, and Lehman Brothers, speculators, high-risk offshore hedge funds bet on the future grain price with no need to take delivery of actual wheat or corn at the end of the contract. Food was no longer a physical "good" to feed the world, but rather a virtual and unlimited trading instrument. The corporate empires expanded as never before, making them rich beyond their wildest dreams. Grain was now entirely decoupled from everyday supply and demand – sending shock waves throughout the world.

The Food and Agriculture Organization of the UN estimates that since 2004, world food prices on average have soared by an unprecedented 240%, sending price-shock waves throughout the global food production and delivery system.

The market was no longer run by people involved in the food business, but by the new casino masters of grain supplies – from Wall Street to London and beyond. The dotcom bubble followed the real estate bubble and we may see a food bubble. Hundreds of millions have already starved to death, mostly in the third world.

They were the greatest of the unwashed, the useless eaters, dirty, diseased, bare-bellied African, Asians and South Americans – a blip on the mainstream corporate networks' map. An unimportant statistics for most, but watched by the most astute movers and shakers of the imperium. To them, that billion-strong army of the walking dead was a key indicator that their master plan of world depopulation was right on schedule.

Food has "become just another commodity like oil or tin or silver whose scarcity and price can ultimately be controlled by a small group of powerful trading insiders."[32]

There is an intricate link between the people who control food and the people who control medicine. It is another aspect of the elite's control of the population.

Rockefeller Global Medical Cartel

Around 1908, the Rockefeller family decided they could make money in the emerging pharmaceutical industry. But, why stop there? Why not turn the practice of medicine, as a whole in the United States, into pharmaceutical medicine based on the petrochemical industry. It seemed logical, as the Rockefellers were already in the oil business.

The family commissioned a study called the Flexner Report, whose findings destroyed all traditional forms of medicine that was practiced at that time in the United States.

The study claimed that traditional forms of medicine were worthless and not scientific. It was a commercial attack against all forms of traditional medicine, and looked to replace them with the Rockefeller pharmaceutical (allopathic) medicine. It was a fully-fledged takeover of the medical system.

Over the years, it has grown to the current point in which under the new national health care law, the government has launched a surreptitious program to map out all treatments and all diagnoses allowed for all diseases and disorders. If you're a part of the national health system, you could be forced to go through certain procedures.

The authorities will make a diagnosis, and you have to take the appropriate medicine. If you reject or do not agree to take the prescribed medicine, they can put small transmitters in pills to see that you take them. This is what one future looks like – A Brave New World. We will find ourselves locked into a health-care system, inaugurated by the Rockefellers in the United States and spread worldwide, which allows only one type of medical practice?

Cradle-to-Grave Mentality

The drug cartel consists of a group of institutions and organizations, some state-run, some not, which is trying to impose a system of cradle-to-the-grave medical-care on all of mankind. It may sound good, but what it really means is that we can be forced throughout our lives to take toxic and destructive drugs. The effect of these drugs may severely weaken the population, just as it appears is happening in America, today.

Even unborn children in the industrialized world are now considered and labeled as medical patients, and are included in the health system from birth. The highly toxic Hepatitis B vaccine is currently given to all children in the United States before they leave the hospital after birth, unless parents explicitly reject it.

Direct evidence shows that the US health care system kills 225,000 people per year, of which 106,000 died because of drugs approved by the FDA. This means that in one decade, 2.25 million people died due to the shortcomings in US health care, as stated in a July 26, 2000 *Journal of the American Medical Association* study called, "Is U.S. health really the best in the world?" The article was written by Dr. Barbara Starfield who was working at the academy for public health at Johns Hopkins.

In other words, because the FDA approves toxic drugs, then they are complicit and responsible before and after the events. Sadly it seems that FDA is not an organization whose task is to protect the patient. Pharmaceutical companies, it appears, are FDA's clients. Why would the FDA approve dangerous drugs? Because the best customers of the FDA are the drug companies. But ultimately, it comes down to that same kind of callous indifference to human life as practiced by the Third Reich.

The overall objective of the pharmaceutical drug cartel is population control. In other words, undermine the ability of people to think, to feel, to have life experiences, to understand what is being done and who does it. This is the perfect system if you are planning to control the world. Cradle-to-grave health care with toxic drugs that destroy you as people.

The goal of globalization is a global management system. Politically speaking, this is what some call the "new world order." The dilemma is how do you sell such evil to sane, articulate, intelligent, attentive and independent people? You don't.

What is needed is people that are as weak as humanly possible, and it turns out that a most effective way to achieve this is by massive drugging of the population. This is being done through pharmaceutical medicine. This is exactly what is happening worldwide, and there are statistics to prove it.

If we add to the death toll the number of people who are affected or suffering from serious adverse reactions to certain drugs, we are talking about between thirty and forty million people. These people cannot think, cannot function independently, and *they can be easily convinced to follow orders*. This is the bottom line of the operation.

Africa

In the mid-seventies, the World Health Organization, which is a part of the UN apparatus, announced that it had eliminated smallpox in Africa. It was an occasions to celebrate with champagne, the greatest medical victory in the history of mankind. Then, almost ten years later, investigative journalist Jon Rappoport obtained exclusive information from an absolutely reliable source, about a secret meeting in Geneva, which came shortly after the WHO announced their victory. At this meeting it was decided, that the smallpox vaccine would be shelved and never used again.

Why? Because this highly dangerous vaccine itself was causing smallpox. In fact, this has been an open secret for decades. Doctors know, researchers know, Robert Gallo from the National Cancer Institute knew. Gallo was the one who said that if given a vaccine, like smallpox, to populations whose immune system had been suppressed, you risk killing huge numbers of people. That's exactly what happened. Thus, the World Health Organization had no choice but to shelve this vaccine.

But we are still left with an obvious question: What about WHO's claim that smallpox had been eradicated in Africa? True, the visible signs of smallpox such as skin lesions were no longer visible on the millions of people who were given the smallpox vaccine. But then people began to develop diseases unknown at that point in the Third World.

Soon after, people in Africa were dying by the tens of thousands. Cases of meningitis became common. To avoid worldwide indignation and a scandal, the WHO and the United Nations simply changed the diagnosis, making it possible for them to claim victory in their fight against smallpox. People with smallpox were now being diagnosed with AIDS. This was a cover up of a gigantic crime perpetrated by the World Health Organization in the Third World and particularly in Africa.

Order out of Chaos

We are witnessing "controlled chaos." The elite are afraid of uncontrolled outwardly manifestations of violence from the people – unless the violence itself is managed by the elite. The type of chaos they are looking for occurs in the human body. In other words, it is the chaos of being poisoned. If you introduce harmful, toxic drugs into your body throughout your life, your body will begin to breakdown at some point. Chaos. The chaos of being intoxicated – literally poisoned. But, in this case, the poisoning will be internal, and to some extent controlled by drugs such as sedatives or tranquilizers.

We see millions or perhaps even billions of people, everyday, who are nothing more than the shell of what they once were. They're walking dead – they are submissive and obedient.

Please understand, world government and the medical cartel go hand in hand. They are the same thing because at the highest levels of power, the elite know that the final goal of the pharmaceutical cartel is to destroy, weaken and shorten human life.

As I have noted before, the globalists want weak willed populations. If you are the elite, how do you get there? The pharmaceutical industry.

This is a hidden part of the operation. We have to ask yourselves: What are the key weapons of the global state? How do they control the minds, bodies and people? Again, please understand that the global police state and the medical cartel work together towards the same final objective – total enslavement of the people.

There is another aspect of this secret agenda we need to look at.

Codex Alimentarius – Depopulation by Stealth

Created by the UN in 1961, with the mandate of guarding the health of consumers, Codex Alimentarius (Latin for *food code*) is an industry dominated regulation authority as well as an UN-sponsored organization, which – under the auspices of the World Health Organization (WHO) and the Food and Agriculture Organization (FAO) – established global trade standards for foods. In principle, it has no legal standing but the Codex has risen to the level of de facto legal standing because Codex Alimentarius is administered by the WHO and FAO.

In 1994, the World Trade Organization (WTO) replaced the General Agreement on Tariffs and Trade (GATT) and was given the trade-sanction *power to enforce* the Codex, and other standards and guidelines, as a means of harmonizing food standards globally for easy trade between countries. In other words, once Codex guidelines, rules and regulations become ratified and approved, they become mandatory for any member-country of the WHO.

International organizations, which are concerned for our well-being and worried about the nutrition and diet of humanity sounds great in theory. Except that Codex has little to do with saving lives and has all to do with the elites' plans to depopulate the planet.

The recently approved German Plan of Codex states that *no* vitamin, herb, or mineral can be sold for preventive or therapeutic purposes.

> Potency levels of supplements would be severely limited. All supplements sold to or in member nations of the Codex would have to be approved according to these draconian guidelines.[33]

Why is this being done? Because nutrients make people smarter, stronger, and more free and independent.

Please understand that Codex is the enemy of everyone except those who will profit from it. Plus there is a hands-on, direct association with those who committed horrendous crimes against humanity during the Nazi regime:

> One of those found guilty was the president of the megalithic corporation I.G. Farben, Hermann Schmitz. His company was the largest chemical manufacturing enterprise in the world, and had extraordinary political and economic power and influence with the Hitlerian Nazi state. Farben produced the gas used in the Nazi gas chambers, and the steal for the railroads built to transport people to their deaths.[34]

What does it have to do with Codex? The corporations and the people behind the initiative are the same companies that supported Nazi Germany: BASF, Hoechst, and Bayer.

> They are three of the eight largest pesticide companies in the world. They are pharmaceutical companies. They are chemical companies. They are genetically modified food companies. They are the three basic companies that once

made up the Nazi chemical cartel I.G. Farben. Farben built Auschwitz, the concentration camp.[35]

Farben helped put Hitler over the top and into the role as the leader of Germany. Is there an obvious connection here or am I imaging things?

> While serving his prison term, Schmitz looked for an alternative to brute force for controlling people and realized that people could be controlled through their food supply. When he got out of prison, he went to his friends at the United Nations (UN) and laid out a plan to take over the control of food worldwide. A trade commission called Codex Alimentarius was re-created under the guise of it being a consumer protection commission. But Codex was never in the business of protecting people. It has always been about money and profits at the expense of people.[36]

Please understand Codex is a weapon being used to reduce the level of nutrition worldwide. For example, "Codex does not set limits for most of the dangerous industrial chemicals that can be used in food. In 2001, 176 countries including the U.S. got together and decided that twelve highly toxic organic chemicals, out of which nine are pesticides, known as persistent organic pollutants (POPS) were so bad that they had to be banned.

> Under Codex, seven of the nine forbidden POPS will again be allowed in the production of food. All together, Codex allows over 3,275 different pesticides, including those that are suspected carcinogens or endocrine disrupters. There is no consideration of the long-term effects of exposure to mixtures of pesticide residues in food.[37]

Surprised? You shouldn't be. The war being waged worldwide today and tomorrow is not against terror but rather against these pesky creatures known as humans. According to the projections of the WHO and the FAO, a minimum of 3 billion people will die from the Codex mandated vitamin and mineral guideline alone.

> Under Codex, every dairy animal can be treated with growth hormone, and all animals in the food chain will be treated

with sub-clinical levels of antibiotics. Codex will lead to the required irradiation of all foods with the exception of those grown locally and sold raw. These food regulations are in fact the legalization of mandated toxicity and under-nutrition. The WHO and FAO estimate that of the three billion people initially expected to die as the result of the Codex vitamin and mineral guidelines, two billion of them will die from the preventable diseases that result from under-nutrition, such as cancer, cardiovascular disease, diabetes, and many others. Those who will live will be the wealthy elites who are able to somehow provide themselves with sources of clean food and other nutrients.[38]

With advances in science and technology, it was only a matter of time before Big Pharma and the Rockefeller family turned their attention to molecular and synthetic biology and gene technology.

Artificial Life, Self-Replicating Synthetic DNA

Imagine bacteria, fitted with artificial DNA, harnessed to churn out an anti-malaria vaccine – that is happening already in California. Or imagine bacteria with synthetic genes that make them light up when parasites are detected in drinking water – that has been proven to work at Imperial.[39]

On Earth, all life is dependent upon the nucleic acids, DNA and RNA. The DNA is a code for life, the twisting strands that hold the genes of every living thing on Earth, essentially comes down to four basic molecules, a long polymer of sugars, linked together by a phosphate – adenine, cytosine, guanine and thymine, better known by their first letters A, C, G and T. It's the order of these that conveys genetic information.

And because these molecules are well understood, they can be manufactured synthetically and rearranged to design brand new genes.

> The phosphate can be modified with a sulfur atom replacing oxygen and the resulting molecule can still undergo base pairing with normal nucleic acids. These synthetic molecules can actually be used by the normal cellular ma-

chinery if they're supplied to bacteria, creating an expanded genetic code.[40]

Now, in one of the biggest breakthroughs in recent history, scientists have created a synthetic genome that can self-replicate. They have taken a cell and modified the genes of a cell by inserting DNA from another organism. And the bacteria replicated itself, thus creating a second generation of the synthetic DNA.

> The cell is totally derived from a synthetic chromosome, made with four bottles of chemicals on a chemical synthesizer, starting with information in a computer.[41]

The organism will do exactly what the scientist intended: a living thing, but under the control of Man.

> If the 19th Century was all about the revolution of harnessing energy from fossil fuels, and the 20th Century was about exploiting the power of data, this century will be about controlling biology.[42]

What is amazing about this, is that the cell was assembled and sparked into life in a laboratory. This technology has taken mankind across a threshold. A turning point that marked a coming of age of a new science called, synthetic biology, founded on the ambition that one day it will be possible to design and manufacture a human being.

In other words, you can get a DNA of anything here on Earth and create organisms that never before existed entirely from non-living materials.

> Scientists are creating new life forms but at the same time, we are creating life forms that the human immune system and the world have never so far experienced.[43]

As such, it will revitalize perennial questions about the significance of life — what it is, why it is important and what role humans should have in its future.

Synthetic genes have then been inserted into a bacterium, which has had its own original DNA stripped out. Collectively, these DNA/RNA substitutes are called XNAs.

One notable property of XNA molecules is they are not biodegradable. The researchers also have developed enzymes that can synthesize XNA from a DNA template, plus others that can 'reverse transcribe' XNA back into DNA. This means they can store and copy data just as DNA can.

The investigators subjected an XNA molecule to artificial natural selection in the lab by introducing mutations into its genetic code. By allowing the different versions of the molecule to compete against each other for binding to another molecule, the team ended up with a shape that bound tightly and specifically to the target – just as one would expect of DNA under the same conditions. This makes XNA the only known molecules other than DNA and RNA capable of Darwinian evolution. Heredity – information storage and propagation – and evolution, two of the hallmarks of life, can be implemented in polymers other than DNA and RNA."[44]

From Eugenics to Genetics

In reality, synthetic biology is a straight-line extrapolation of molecular biology and just a step away from your good, old eugenics. It all began with the Rockefeller family.

> The genetic engineering initiative of the Rockefeller Foundation was no spur of the moment decision. It was the culmination of the research it had funded since the 1930's. During the late 1930's, as the foundation was still deeply involved in funding eugenics in the Third Reich, it began to recruit chemists and physicists to foster the invention of a new science discipline, which it named molecular biology to differentiate it from classical biology.
>
> The idea had been promoted during the 1920's by Rockefeller Institute for Medical Research biologist Jacques Loeb, who concluded from his experiments, that echinoderm larvae could be chemically stimulated to develop in the absence of fertilization, and that science would eventually come to control the fundamental processes of biology. The people in and around the Rockefeller institutions saw it as the ultimate means of social control and social engineering – eugenics.

Borrowing generously from their work in race eugenics, the foundation scientists developed the idea of molecular biology from the fundamental assumption that almost all human problems could be 'solved' by genetic and chemical manipulations.

The foundation's research goal was to find ways to reduce the infinite complexities of life to simple, deterministic and predictive models. The promoters of the new molecular biology at the foundation were determined to map the structure of the gene and to use that information 'to correct social and moral problems including crime, poverty, hunger and political instability.'[45]

Another key associates working for the Rockefellers, and involved in reducing human problems to a basic common denominator of death was Dr. Franz J. Kallmann, a psychiatric geneticist at the New York State Psychiatric Institute. Dr. Kallmann was also a founding President of a new eugenics front organization, the American Society of Human Genetics which later became a sponsor of the Human Genome Project.

The multibillion-dollar project was, appropriately enough, housed at the same Cold Spring Harbor center that Rockefeller, Harriman and Carnegie had used for their notorious Eugenics Research Office in the 1920's. Genetics, as defined by the Rockefeller Foundation, would constitute the new face of eugenics.

Kallmann was a strong advocate of practicing elimination or forced sterilization on schizophrenics. In 1938,

> ... he demanded the forced sterilization of even healthy offspring of schizophrenic parents to kill the genetic line.[46]

In the 1960s and 70s, while brother John D. Rockefeller III was mapping plans for global depopulation, brothers Nelson and David were busy with the business side of securing the American Century. American agribusiness was to play a decisive role in this project, and the development of genetic biotechnology would bring the different efforts of the family into a coherent plan for global food control in ways simply unimaginable to most.[47]

That was then.

Now, after 10,000 years of natural genetic manipulation by selective breeding, we have finally gained direct access to the genetic code, DNA. We are now doing to genetic engineering what engineers have done since the Stone Age: collect, refine and repackage nature so that it is easier to make new and reliable things.

Welcome to the world of synthetic biology, which in laymen's terms is genetic modification. It is the dawn of a new revolution in molecular biology and genetic engineering.

> Over recent years, Genetic Modification has led to crops that are resistant to weed killers or insecticides. Most startling are the goats that carry the spider gene that produces silk. But what is coming next with synthetic biology takes this research into an entirely different league, and only now is it entering the public consciousness.[48]
>
> It is based on what's called recombinant DNA (rDNA), and it works by genetically introducing foreign DNA into plants to create genetically modified organisms. One way or other, the Rockefeller Foundation aims to reduce population through human reproduction by spreading GMO seeds. It's doing it cooperatively with the UN World Health Organization (WHO) by quietly funding its reproductive health program through the use of an innovative tetanus vaccine. Combined with HCG natural hormones, it's an abortion agent preventing pregnancies, but women getting it aren't told. Neither is anything said about the Pentagon viewing population reduction as a sophisticated form of biological warfare to solve world hunger.[49]

Ultimately, this is about taking control of nature, redesigning it and rebuilding it to serve the whims of the controlling elite. No wonder the phrase "playing God" comes up in almost every conversation. With it comes a grand historical vision.

Now, what will happen if a private trust or the Pentagon comes to possess major cellular breakthrough, which has huge implications for mankind? The Pentagon military industrial complex can spread a virus and use a vaccine to extinguish, what the Pentagon calls, undesirable human behavior. Stuff of science fiction, you say?

There are many disturbing potentials. We can now build viruses from the genomes. We have already been able to create a synthetic poliovirus. And as Britain's *Daily Mail* reported:

> This technique can be used to recreate terrible viruses from our past, like Ebola and a 1918 flu strand that killed up to 40 million people.[50]

What's more, there are easier ways to recreate microbes. You can simply add the right gene to a close relative.

The Future is Now

What about the future? How will this technology be used in the future? It is after all, a double-edged sword. Take children, for example. Not only will parents have an opportunity to genetically select or to elect their children, building their characteristics block by building block, and in the process creating a new breed of children – no disease, level of IQ etc., not to mention mixing in other people's DNA in with the couple's to literally produce a new breed of humans. A question is: If we give our children genes that we don't have, are they still our children?

It can be argued that the child is still you, only the best of you and that a couple could conceive thousands of times without getting the desired results available through selection. Shouldn't the couples choose the best in them to pass on to their children? And if yes, what does the "best" mean? And in what sense the best? How do you define morality? Does it have anything to do with the divine spark of reason? If yes, where does that come from and how can that trait be mapped through DNA? And if it could, would that mean that a synthetic version of it could be controlled through biomedicine?

As a member of the Rockefeller Foundation board and avowed eugenicist, Frederick Osborn said:

> It would in the end be far easier and more sensible to manufacture a complete new man *de novo*, out of appropriately chosen raw materials, than to try to refashion into human form those pitiful relics which remained.[51]

I repeat, the promoters of the new molecular biology at the Rockefeller Foundation are determined to *correct* social and moral problems including crime, poverty, hunger and political instability – all done in the name of population control and population reduction.

But, there is even a far darker element to consider. Parents who misbehave, criminals, dissidents, people who think differently from the official party line of the One World State could have their children's DNA altered during pregnancy as punishment for their non-obedience.

What's more, by altering DNA, the private corporations and governments can create a society without memory: people, whose life's experiences are stored on a memory drive running on a one week cycle through a modified DNA, over and over and over again.

Are we willing to go this route? Why do I ask? Because then, the Bilderberg-controlled bio-companies would hold the key to life in every aspect of your existence. The big question is can we trust the major corporations to do the right thing? Do scientists have morals? What about the corporations? And what about loyalty? Remember, these mega-corporations historically have held no loyalty to any nation, state or God.

Can we trust them with creating life? This is a very important question to ask, because for the first time in history – science can create life. We are giving birth on daily basis to new astounding technologies.

This is not new. In the past 100 years or so, governments began working on bio-weapons – Imperial Japan, Germany, UK, France. And many programs have delved into race specific bio-weapons, ones that would eradicate undesired people.

The problem with the research is that governments, when working on bio-weapons, generally first try to take zoological bio-weapons like Ebola that affected apes, and cross it over to humans. There is evidence that this is what happened with Ebola.

Currently, we have human-cross-species clones growing inside of cow's uterus so they can harvest the organs. They are mixing animal life and human. This will allow cross-species diseases that can then only be eradicated through the Bilderberg controlled organizations, through human experiments, DNA sampling, or using synthetic DNA to implant in human beings.

People or robots, or is there a new species being created? What about rights, human rights? But then, because they are not human, it can be argued that the rights these new organisms have will be more related to animal rights than human rights.

What does the *Strategic Trends 2007-2036* report have to say on the subject? I quote from the report:

> … a more permissive R&D environment could accelerate the decline of ethical constraints and restraints. The speed of technological and cultural change could overwhelm society's ability to absorb the ethical implications… The nearest approximation to an ethical framework could become a form of secular unilateralism, in an otherwise amoral scientific culture.

The ultimate form of social Darwinism will be welcomed at last. Francis Galton's religion will reign supreme, as younger generations will make eugenics a normal part of their life.

Add to it financial destruction, displacement of people across the world, food crisis, wars, famine and decease and you have a perfect storm of depopulation. Again, I quote the *Strategic Trends Report*,

> Declining youth populations in Western societies could become increasingly dissatisfied with their economically burdensome 'baby boomer elders.…' Aging populations, an increase in medical demand and patient expectations are likely to lead to an unsustainable drain on some state health resources and to impact on economic prosperity.… Resentful at a generation whose values appear to be out of step with tightening resource constraints, the young might open the way to policies that permit euthanasia as a means to reduce the burden of care for the elderly. This could lead to a civil renaissance with strict penalties for those failing to fulfill their social obligations.[52]

What they do mean by imposing "strict penalties for those failing to fulfill their social obligations?" It means genocide. It means that in the name of social equality, the elderly will be killed to give way for the younger generation. The *Strategic Trends Report* makes it very clear:

> It will also open the way to policies which permit euthanasia as a means to reduce the burden of care for the elderly.

And there you have it.

This is Transhumanism in its purest and most evil form as envisioned by Aldous Huxley.

In *Brave New World,* Huxley centers on the scientific methodology for keeping all populations outside the elite minority in a permanently autistic-like condition actually in love with their servitude and producing dictatorship without tears.

In a 1961 speech on the U.S. State Department's Voice of America, Huxley spoke of a world of pharmacologically manipulated slaves, living in a "concentration camp of the mind," enhanced by propaganda and psychotropic drugs, learning to "love their servitude," and abandoning all will to resist. "This," Huxley concluded, "is the final revolution."

And thus, we're now at a point, where either the people win, and restore the kind of government we require, in various nations, and among nations. Or this world is going to go into Hell, because the crisis won't quit.

> The people will die of hunger, they will die in increasing numbers; they will kill for food. The structure of society will be destroyed in the fight over food, which is not there. And therefore, either we win this fight against this evil, or there won't be anything to fight for.[53]

Endnotes

1   F. William Engdahl, "Getting used to Life without Food: Wall Street, BP, bio-ethanol and the death of millions," globalresearch.
2   Ibid.
3   Stephen Lendman, *Seeds Of Destruction, F. William Engdahl, Review,* January 22, 2008.
4   Ibid.

5   Richard Freeman, "The Windsors' Global Food Cartel: Instrument for Starvation," *EIR*, December 8, 1995 http://www.larouchepub.com/other/1995/2249_windsor_food.html.

6   DCDC *Strategic Trends Report*, p 9.

7   Marcia Merry Baker, "Food cartels: Will there be bacon to bring home?" *EIR* Volume 26, Number 27, July 2, 1999.

8   "The True Story Behind the Fall of the House of Windsor," *EIR Special Report*, September 1997.

9   "The Food Crisis: Who Shall Rule?" – Editorial, *Campaigner* magazine, November 1973.

10 Richard Freeman, "The Windsors' Global Food Cartel: Instrument for Starvation," *EIR*, December 8, 1995.

11 Marcia Merry Baker, "Food cartels: Will there be bacon to bring home?" *EIR* Volume 26, Number 27, July 2, 1999.

12 Richard Freeman, "The Windsors' Global Food Cartel: Instrument for Starvation," *EIR*, December 8, 1995.

13 William Engdahl, *Seeds of Destruction – The Hidden Agenda of Genetic Manipulation*, pp.138-139, Global Research, 2007

14 Richard Freeman, "The Windsors' Global Food Cartel: Instrument for Starvation," *EIR*, December 8, 1995.

15 Marcia Merry Baker, "To Defeat Famine: Kill the WTO," *EIR*, February 2008.

16 *Keiser Report: Semaphore of Fraud*, August 9, 2012.

17 Rockefeller speech July 1972, Pocantino Hills, NY.

18 "Agribusiness Giants seek to gain Worldwide Control over our Food Supply," Stephen Lendman review of F. William Engdahl's *Seeds of Destruction*, January 7, 2008 Globalresearch.

19 http://en.wikipedia.org/wiki/Green_revolution.

20 Stephen Lendman, "Agribusiness Giants seek to gain Worldwide Control over our Food Supply," review of F. William Engdahl's *Seeds of Destruction*, January 7, 2008 Globalresearch.

21 Ibid.

22 Jeffrey Smith, *Huffington Post*, August 9, 2010.

23 "Potential Health Hazards of Genetically Engineered Foods," Stephen Lendman, February 22, 2008

24 "A Comparison of the Effects of Three GM Corn Varieties on Mammalian Health."

25 http://www.biolsci.org/v05p0706.htm.

26 Stephen Lendman, "Agribusiness Giants seek to gain Worldwide Control over our Food Supply," review of F. William Engdahl's *Seeds of Destruction*, January 7, 2008 Globalresearch.

27 Ibid.

28 http://naturalliving360.com/gm-food-dangers-include-low-fertility-organ-damage-and-hormone-disruption.html.

29   "Agribusiness Giants seek to gain Worldwide Control over our Food Supply," Stephen Lendman review of F. William Engdahl's *Seeds of Destruction*, January 7, 2008 Globalresearch.

30   F. William Engdahl, "Getting used to Life without Food: Wall Street, BP, bio-ethanol and the death of millions," globalresearch.

31   Ibid.

32   Ibid.

33   Scott Tips, "A Meeting of Two," *Health Freedom News Board*, December 2004.

34   Barbara Minton, "Billions of People Expected to Die Under Current Codex Alimentarius Guidelines," *Natural News,* July21, 2009.

35   http://www.buildfreedom.com/news/archive.php?id=549.

36   Barbara Minton, "Billions of People Expected to Die Under Current Codex Alimentarius Guidelines," *Natural News,* July21, 2009.

37   Ibid.

38   Ibid.

39   http://www.bbc.co.uk/news/science-environment-17436365.

40   http://arstechnica.com/science/2012/04/synthetic-dna-substitute-gets-its-own-enzymes-undergoes-evolution/.

41   "Synthetic DNA Breakthrough we Now Create Artificial Life," http://m.io9.com/5543843/scientists-create-artificial-life-+-synthetic-dna-th...

42   David Shukman, "The strange new craft of making life from scratch," March 26, 2012, *BBC Science & Environment.*

43   http://www.lifeslittlemysteries.com/830-whats-synthetic-biology.html.

44   Charles Q. Choi, "XNA, Synthetic DNA, Could Lead To New Life Forms, Scientists Say," Huffington Post, April 19, 2012

45   "Seeds of destruction – The Hidden Agenda of Genetic Manipulation, William Engdahl, pp.138-139, Global Research, 2007, p.155-6.

46   Ibid, p. 91-92.

47   Ibid, p.94-95.

48   David Shukman, "The strange new craft of making life from scratch," March 26, 2012, *BBC Science & Environment.*

49   "Agribusiness Giants seek to gain Worldwide Control over our Food Supply," Stephen Lendman review of F. William Engdahl's *Seeds of Destruction*, January 7, 2008 Globalresearch.

50   John Naish, "The Armageddon virus: Why experts fear a disease that leaps from animals to humans could devastate mankind in the next five years," *Daily Mail*, October 14, 2012

51   Frederick Osborn, *The Future of Human Heredity: An Introduction to Eugenics in Modern Society*, Weybright and Talley, New York, 1968, pp. 93-104.

52   DCDC *Strategic Trends Report*, p.79

53   Marcia Merry Baker, "World Food Shortage, A British Policy Success," *EIR*, April 16, 2010

Chapter Three

Programming the Masses

Mass media is designed to reach the largest audience possible, it include television, movies, radio, newspapers, magazines, books, records, video games and the Internet. Today, a Bilderberg-dominated, highly centralized group of powerful individuals and multinational companies own the world news media.

Media control and manipulation does not mean that a monolithic "Big Brother" serves up one, and only one distorted version of reality. The media cabal defines the parameters of what is "news," and, therefore, shapes the latitude of the public discourse and the range of policy options available for debate. It continually disinforms, misinforms, and dumbs-down the populations.

What's imperative to understand is that no matter which media outlet you read, watch, surf or listen to, the entire mainstream and most alternative news sources are controlled by the same interlocking Octopus.

Mass Psychology Model

Today's elite have a precedent: the pagan cult ceremonies of the decadent Egyptian and Roman Empires. And these controlling entities have a history all their own. The continuity of the cult of Apollo is important to understand.

There are "black nobility" families whose families conjoined political traditions trace back to the Roman empire.

> The republic and empire under which their ancestors lived was in turn controlled by Rome branch of the cult of Apollo. That cult, was during that time, variously, the chief usurious debt-farming institution of the Mediterranean region.[1]

The cult was a political intelligence service, and the creator of other cults.

From the death of Alexander the Great until the cult of Apollo dissolved itself into the cult of Stoicism, creating during the second century B.C., the base of the cult was Ptolemaic Egypt, from which the cult controlled Rome. In Egypt, the cult of Apollo syncretized the cult of Isis and Osiris into a direct imitation of the Phrygian cult of Dionysus and its Roman imitation, the cult of Bacchus.

It was there that the cult of Stoic irrationalism was created by the cult of Apollo. It was the cult of Apollo that created the Roman Empire, which then created the Roman law on the basis of the antihumanistic Aristotelean Nicomachean Ethics. That is the tradition, which the old "black" Roman families transmit. These Roman families, over time, became known as the "Venetian Black Nobility." Today these same families hold key positions within the inner circle of organizations such as the Bilderberg group.

This tradition has persisted under various institutional covers, always preserving its essential world-outlook and doctrine intact. The British monarchy, the aristocratic landlord class of Europe, and the European-dominated, feudalist factions of the Maltese Order, are the modern, concentrated expression of the unbroken tradition and policies of the ancient Apollo cult.

An Aristotelean knows that:

> … generalized scientific and technological progress, given the conditions of education and liberty of innovation which progress demands, produces in the citizen a dedication to the creative potential of the human mind which is antithetical to the oligarchic system.
>
> What the Aristoteleans have hated and feared down through the millennia is their knowledge that persistent, generalized scientific and technological progress, as the guiding policy of society, means a republican hegemony which ends forever the possibility of establishing oligarchic world-rule.[2]

Today's elite have resorted to the same methods used by the ancient priests of Apollo – the promotion of Dionysian cults of drug-cultures, orgiastic-erotic countercultures, deranged mobs of "machine-breakers" and terrorist maniacs. They use psychological warfare to turn the combined force of a demented rabble against those forces in society that are dedicated to scientific and technological progress.

And no better way exists than through the mass brainwashing of the population, and the shortest way to the hearts and minds of the people is through the all-encompassing, global communication.

Television

The greatest form of control is when you can be manipulated the population into believing that they are free, while they are being dictated to and manipulated. One form of dictatorship is being in a prison cell, seeing the bars. The other, a far more subtle form of dictatorship, where you can't see the bars, but think you are free. The biggest hypnotist in the world is that oblong box in the corner of the room that tells people what to believe. Television, with its reach into everyone's home, creates the basis for the mass brainwashing of citizens.

> Television causes people to suspend their critical judgment capabilities because the combination of sound and images places the individual in a dream-like state, which limits cognitive powers.[3]

Hal Becker from the Futures Group contends that through the control of television-news programming, he can create popular opinion by manipulating the way you think and act.

> Americans think they are governed by some bureaucrats in Washington who make laws and hand out money. How wrong they are. Americans are ruled by their prejudices and their prejudices are organized by public opinion.… We think that we make up our minds about everything. We are so conceited. Public opinion makes up our minds. It works our herd instinct, like we are frightened animals.

But there is a very big difference between beast and man. A major difference is in our search for eternal Truth – for the meaning of Life. Truth lies in a higher order of processes and in the creative powers of the individual human mind. It's a moral problem, and *a problem of mankind's destiny*, one, which no animal can ever solve.

Every generation must advance beyond that of the preceding generation. And that hope – that it will happen – should be what's on the mind of a person dying of old age: that their life has meant something, and has helped to create a better world than they knew.

Polling by Numbers

Stereotypes are created, and manipulated, by the gurus of mass communication and psychological warfare. The idea is not to make you think too clearly, or too profoundly about the images you receive, but instead to react in a Pavlovian manner, to the stimuli provided.

Edward Bernays, Freud's nephew and one of the founders of public opinion manipulation techniques stated that:

> We are governed, our minds are molded, our tastes formed, our ideas suggested, largely by men we have never heard of. Whatever attitude one chooses to take toward this condition, it remains a fact that in almost every act of our daily lives, whether in the sphere of politics or business, our social conduct or our ethical thinking, we are dominated by a relatively small number of persons, a trifling fraction of our hundred and twenty million [U.S. citizens at the time], who understand the mental processes and social patterns of the masses. It is they who pull the wires, which control the public mind, and who harness old social forces and contrive new ways to bind and guide the world.[4]

In this manner, irrationality has been elevated to a high level of public consciousness. The manipulators then play upon this irrationality to undermine and distract the grasp of reality governing any given situation. And as the problems of a modern industrial society become more complex, the easier it is to bring greater and greater distractions to bear, so that what we ended up with is the absolutely inconsequential opinions of masses of people, created by skilled manipulators. These "opinions" are then assumed as representatives of "scientific" facts.

We are talking about Freudian mass psychology and its appeal to infantilist, animal-like behavior, the touchy-feely thing, designed to bypass the creative reasoning powers of individuals, informed by moral judgment and the eternal search for universal Truth.

In fact, when it comes to television, the issue of truth has never been an issue. Television is not the truth. Television is an amusement park, a group of jugglers, belly dancers, storytellers, singers and stripers. But we, the people, have been completely hypnotized by the Tube. You sit there; day after day, night after night – TV is most of what you know! Five percent of Americans read more than five books per year (around 15 million folk), yet one billion people watched the Oscar awards. You dream like a tube, speak like a tube, smell, dress, act like your television. Most people feel they are better acquainted with Paris Hilton, Britney Spears or Lady Gaga than with their own husband or wife. This is madness!

How many millions of you are prepared to believe anything the tube tells you? What's more, the people at the top are prepared to tell you anything in the name of "war against terror," audience share and advertising dollars, as long as you vote for them, buy their products and allow them to brainwash you.

> Television provided an ideal means to create a homogenous culture, a mass culture, through which popular opinion could be shaped and controlled, so that everyone in the country would think the same.[5]

Again, television is not about the "truth," it is about creating a reality. It does not matter one bit whether the images you see on television are real or copied and pasted from past events, because people believe them to be real, immediate and thus true.

For example, during the March 2010 earthquake in Japan, the mass media showed images of empty supermarket shelves, and stated brazenly that Japan was undergoing the worst water rationing since World War II.

However, the images of empty shelves were taken from stock photography and had little to do with the earthquake or lack of bottled water. In this way, reality, as conveyed by the nightly news, obliterates Truth every night. Emery and Trist indicated that:

> ...the more a person watches television, the less he understands, the more he accepts, the more he becomes dissociated from his own thought process ... television is much more magical than any other consumer product because it makes things normal, it packages and homogenizes fragmentary aspects of reality. It constructs an acceptable reality (the myth) out of largely unacceptable ingredients. To confront the myth would be to admit that one was ineffective, isolated and incapable.... It (television images) becomes and is the truth.

The brainwashers-in-charge of this societal transformation have pulled off the ultimate trick: They have been able to persuade us that what we have been shown is all that there is to see. Subsequently, people will laugh in your face when you try to explain to them a bigger picture and the unseen reality behind the curtain.

What became obvious to the brainwashers is that an appeal to emotionalism was needed in order to break through the population's moral compass – society needed to be reduced to infantilism.

Writing in 1972, Tavistock's leading media expert, Dr. Fred Emery, reported on television's impact on Americans:

> We are suggesting that television evokes a basic assumption of dependency. It must evoke (this) because it is essentially an emotional and irrational activity ... television is the non-stop leader who provides nourishment and protection.

In an informative report on television's impact on the cognitive powers of an individual, investigative journalist, Lonnie Wolfe says that both Emery and Eric Trist, who until his death in 1993 headed Tavistock's operations in the United States, noted that:

> ... all television had a dissociative effect on mental capabilities, making people less able to think rationally. Viewers, as they become habituated to watching six hours or more of television daily, surrender their powers of reason to the images and sound coming from the tube.[6]

Tavistock recognized that habituated television watching destroys the ability of a person for critical cognitive activity. In other words, it makes you stupid.

To bring society down to the level of a beast is especially important from the point of view of Tavistock, especially if they are to control the planet Earth.

> Since the only source of increase of mankind's power, as a species, within and upon the universe, is that manifold of validated discoveries of physical principle, it follows, that the only form of human action that distinguished man from beast, is that form of action, which is identified as cognition, by means of which the act of discovery of accumulated verifiable universal physical principles is generated. It is the accumulation of such knowledge for practice, in this way, from generation to generation, which defines the provable evidence of the absolute difference between man and beast.[7]

This alleged myth is at the heart of the control mechanism the elite are using to manipulate the people. Most Americans and Europeans believe there is such a thing as a free press. This is one of the key areas in brainwashing a population. And most Americans and Europeans get majority of their news from state-controlled television, under the misconception that reporters are meant to serve the public by offering the people critical choices and distinct points of view.

In fact, reporters do not serve the public and the elite controls all the points of view. Reporters are paid employees and serve the media owners, whose company's shares are traded on Wall Street. In other words, the people who own the media tell us what to think.

By controlling the extremes of this system, the elite have manipulated people into believing that the choices we make are independent and are based on access to critical information. When in fact, the information we receive through mainstream corporate media is part of a manipulation and control.

For example, what do international terrorism; the world's financial markets, empire builders and capitalism have in common? Their utter dependence on drug profits for their very existence.

The war on drugs is a sham. In researching global cash flows, it is staggering to find out that the amount of profit generated annually by the drug trade is close to a trillion dollars per year *in cash*.

Drug money is now an essential part of the world's banking and financial system because it provides the liquid cash necessary to make the monthly payments on the huge stock and derivate investment bubbles. However, you never hear this from the corporate media because it is run by the same financial and political interests whom control and profit from the nefarious drug business.

> American media elites practice a brutal, albeit well-concealed, form of 'wartime' news censorship, but the mechanisms of this control are now openly acknowledged. John Chancellor, the longtime NBC-TV news anchorman, in his autobiographical account of life in the news room, *The New News Business*,[8] admitted that, through formal structures such as the Associated Press, informal 'clubs' such as the New York Council on Foreign Relations, decisions are made, on a daily or weekly basis, about what the American people will be told, and what stories will never see the light of day.[9]

The Council on Foreign Relations is a US arm of the powerful and secretive Bilderberg Group. It is also the premier US foreign policy think tank in the United States, and is one of the central institutions for socializing American elites from all major sectors of society. It is where they work together to construct a consensus on major issues related to American imperial interests around the world. As such, the CFR often sets the strategy for American policy, and wields enormous influence within policy circles, where key players often and almost always come from the rank and file of the CFR itself.

America's corporate media is an integral part of the economic establishment, with links to Wall Street, the Washington think tanks, Club Bilderberg and the Council on Foreign Relations (CFR) and through them to the world's top brainwashing center, Tavistock Institute. CFR is:

> … the premier U.S. foreign policy think tank in the United States, and is one of the central institutions for socializing American elites from all major sectors of society (media,

banking, academia, military, intelligence, diplomacy, corporations, NGOs, civil society, etc.), where they work together to construct a consensus on major issues related to American imperial interests around the world. As such, the CFR often sets the strategy for American policy, and wields enormous influence within policy circles, where key players often and almost always come from the rank and file of the CFR itself.[10]

In a October 30, 1993 "Ruling Class Journalists" essay, *Washington Post* ombudsman Richard Harwood candidly discussed how powerful, private and semi secret organizations such as the Bilderberg Group dominates the news media: "The editorial page editor, deputy editorial page editor, executive editor, managing editor, foreign editor, national affairs editor, business and financial editor and various writers as well as (the now deceased) Katherine Graham, the paper's principal owner, represent the *Washington Post* in the council's membership," observed Harwood. These media heavyweights "do not merely analyze and interpret foreign policy for the United States; they help make it," he concluded.

Rather than offering an independent perspective on our rulers' actions, the Establishment media act as the ruling elite's voice – conditioning the public to accept, and even embrace "insider" designs that otherwise might not be politically attainable.

Social Media – The Mighty Wurlitzer

Almost simultaneously, these perspectives presented in the form of popular opinion are echoed through the Social Media Networks, giving a greater and greater propagation to the disinformation, misinformation, and distorted reality which then spreads like wildfire through Twitter, Facebook, MySpace, Pinterest, Flickr, Digg, Technorati, Messenger, Tweetpeek, Ning, LinkedIn, and endless others.

Think about the power of the Internet.

In one day on the Internet, enough information is consumed to fill 168 million DVDs, 294 billion emails are sent, 2 million blog posts are written (enough posts to fill *Time* magazine for 770 million years), 172 million people visit Facebook, 40 million visit Twitter, 22 million visit LinkedIn, 20 million visit Google+, 17 million visit Pinterest, 4.7 billion minutes are spent on Facebook, 250 million photos are uploaded, 22 million hours of TV and movies are watched on Netflix, 864,000 hours of video are uploaded to YouTube, more than 35 million applications are downloaded and more iPhones are sold than people are born.[11] All in a 24-hour span.

Do you want to know how to make the world believe anything? Through media control. You put something on the television and on the Internet, and it becomes reality. You transmit anything through the social media and it too, becomes reality, through the power of repetition. And people start trying to change the world to make it like the TV set images and social media gossip.

Just look at the statistics.

Forty-two percent of Americans watch TV while they're on their laptops, smart phones or tablets. 31% of those aged 50+ are talking "TV" on social media, 27% 25-35 and 12% up to 18 years. 77% of social network users tweet to tell friends what they're watching.[12] Every month the online population spends equivalent of 4 million years online. These are not empty numbers. These numbers represent people. And people represent access to unbelievable amount of information about their lives, their likes, tastes, dislikes, prejudices and tendencies. These are catalogued, analyzed, adjusted to the needs of the elite and presented in ready-made formats for consumption in ways that project the elite's point of view. This is how they brainwash you, every minute of every day through their ubiquitous control of every media channel available.

This is the real meaning of "mass audience." The concept behind it was the same as discussed by Freud in *Mass Psychology*, that is:

> … individuals participating in the mass phenomenon are susceptible to suggestion, to losing his moral conscience – thus become overwhelmed by the mass experience.

An individual can be made to transfer his or her identity to the group, wherein they become subjected to the most intense forms of suggestion. Providing that the individual's inner sense of real identity is destroyed, he can be manipulated like a child. And as Lord Bertrand Russell wrote in his 1951 book, *The Impact of Science on Society*:

> The social psychologists of the future will have a number of classes of school children on whom they will try different methods of producing an unshakable conviction that snow is black.[13]

Today, with almost 70% of the world adult population using social media, it has becoming easier than ever before. Out of these 70%, 4 out of 5 use smart phones to connect to social media. By the year 2015, the number of smart phones in use will reach one billion units.

This billion will be dominated by Apple, Google and Microsoft, who will enjoy 90% of market share with their respective platforms. These are Bilderberg-run companies. A billion smart phones connected to the happy world of ever after. And this liberating experience comes at a price. It is called surveillance. Yes, you are being watched, listened to, and profiled and catalogued 24/7.[14]

One Happy Family

In fact, cell phones are one of the three most important breakthroughs in the surveillance game. The others are GPS and the ability to watch us. The cellular telephone has become an extension of our bodies 24/7, which means, your whereabouts are known at all time. Also, there is Skyhook, "the fastest, most accurate, most reliable and most flexible location system on the market today."[15]

People with smart phones with GPS, Google employees in special vehicles moving around the world are recording the coordinates of all WI-FI. Each router and its location are being recorded!

This is but the tip of the proverbial iceberg. New search capabilities and indexes, including translation, speech recognition technology, text recognition in videos are some of the latest innovation which have changed the rules of the game – smart cameras with block recognition of numbers, letters and faces linked to massive high-speed databases put everyone on the grid.

The main thing that has changed – nothing is removed, nothing is erased, and nothing is forgotten. Everything you've ever done; every stupid picture you have posted, every nasty blog on a bad day and off-color joke, every text – everything is kept forever. Forever!

It is stored, and not just stored; it is indexed and linked – tied to your name forever. A profile is created of you in a database available to the government and the intelligence community. And you haven't even done anything!

But that's not the point. The point is that your privacy has been violated forever in the name of whatever "-ism" that is in vogue at the moment. And if it is not in vogue, don't worry, it will be, just as soon as the government gives it legitimacy and fully promotes it through the Internet and the social networks using their agents of change as the battering ram against anyone who might object.

Total Surveillance Network

A decade ago, the backbone of the Total Surveillance network was an activity called "data mining" or knowledge discovery, which is the automated extraction of hidden predictive information from databases which consisted of massive collection of information about people's lives, from multiple sources – a bewildering plethora of state-of-the-art technology and custom-designed data-gathering software, including RFID microchips, biometrics, DNA chips and implantable GPS chips, keyword search programs that sift through large databases of text-based documents and messages looking for keywords and phrases based on complex algorithmic criteria. Voice recognition programs convert conversations, targeting an individual's voice pattern, into text messages for further analysis. That was 10 years ago. That's an eon in computer terms.

Compare with an early December 2012 proposal from members of the United Nation's International Telecommunications Union (ITU) who have agreed to work towards implementing a standard for the Internet that would allow for eavesdropping on a worldwide scale.

At a conference in Dubai this week, the ITU members decided to adopt the Y.2770 standard for deep packet inspection, a top-secret proposal by way of China that will allow telecom companies across the world to more easily dig through data passed across the Web.[16]

Today, privacy does not exist. Shame does not exist. Anonymously and openly, people all over the world post everything there is about them on the Internet via Facebook, MySpace, Twitter, etc: name, address, phone number, where you went to school, where you work, who your friends are, your attitude towards an endless array of issues, both political and social, what you did last night, last week, or last year, and so on.

Endless quantities of information about you – the Stasi's wet dream come true; and all of it is available on the Web – for free. Why do people do that? Because we're convinced that our Instagram snaps and Twitter quips are pure genius, that our very special self-expression is unique, witty, and brimming with creative value, when in fact, they are boring, predictable for the most part and mostly the illiterate ramblings of a dead brain.

Social media, in fact, has given status to the "celebrity" which few real men attain. And what is a "celebrity" if not the ultimate human pseudo-event, fabricated on purpose to satisfy our exaggerated expectations of human greatness. This is the ultimate success story of the twenty-first century and its pursuit of illusion. A new mold has been made, so that marketable human models – modern 'heroes', can be mass-produced, to satisfy a market, and all without any hitches. The qualities that now commonly make a man or woman into a "nationally advertised" brand are in fact a new category of human emptiness.

The world we have been peering into is somehow beyond good and evil. It is a world of sentimentality, of makeovers, of people who are willing to shed a tear in front of the whole world.

Instead of saying indefensible things and trusting that the audience will love them anyway, they explain their hardships and plead their case in front of the whole world, which is a very unreasonable thing to do. I find it disturbing to see people allowing themselves to be used like Kleenex, but then everybody else in our trash culture appropriates profound concepts for shallow ends. Initial triumphalism giving way to grudged defeatism – is today's human psyche as seen through the prismatic binoculars of an intrusively globalised, theraupeutised and "Coca-colised" world morality.

Twitter – Blue Bird of Happiness

In an insightful piece entitled "Symbolic Literacy," author Michael Tsarion observes that we suffer from "chronic symbol illiteracy" and that we are subjected to subliminal and subtextual persuasion that constitutes what he refers to as a psychic dictatorship.

This dictatorship, Tsarion says, "involves the deliberate and subversive manipulation and public purveyance of words, images, numbers, colors, rhythms, and symbols which are subsequently directed, via ubiquitous media oracles, toward the limbic areas of the human brain," which, he continues, "produces an elaborate and insidious cryptic language specifically designed to stimulate conflict between fantasy and reality."

Symbols are oracular forms, "mysterious patterns creating vortices in the substances of the invisible world. They are centers of a mighty force, figures pregnant with an awful power, which, when properly fashioned, loose fiery whirlwinds upon the earth."[17] How many people have asked themselves what is the dictionary definition of Ground Zero, why Google is called Google and what does the word mean, if anything. Why Twitter is called Twitter and what is the meaning of the cute little blue bird the company uses as it's universally recognized image?

The "Land of Memory" has always been the primary objective of mind control and counter-insurgency operations. There is a phrase which is not used as much these days: "the blue bird of happiness."

What many people do not realize is that this term had its origins in *The Blue Bird* (1909), a most famous work by the Belgian Nobel Prize-winning author and dramatist, Maurice Maeterlinck. In this play, two children set off on a search for the "Blue Bird of Happiness." This search leads them on many adventures – a kind of initiatory quest for the Holy Grail. Many of the motifs of

Maeterlinck's play are repeated in the CIA's search for perfecting mind control, a search that began with "Project Bluebird."

The Land of Memory, of course, was the target of the Bluebird project: to enter that Land, in another person's mind, to go through the drawers, rearranging the furniture, and leave unnoticed. Once the Korean War started, and American POWs began making bizarre, pro-Communist statements after a mysterious sojourn in Manchuria, the world was introduced to the concept of 'brainwashing,' and the Bluebird took on enormous importance.

If the Communists could alter the consciousness of American soldiers, then "war" took on a completely different nature: it became a war of culture against culture, of atheism against religion, of race against race, of Darkness against Light. This was a war not to be fought by bullets alone; psychological warfare operations were ramped up at the same time as Bluebird went into full swing, and what William Sargant would call in 1957, "The Battle for the Mind," had begun.

Through an "innocent" child's tale, brainwashers and counter-insurgency operatives have embarked on a "sacred" quest that would lead them into humanity's deepest secrets; by delving into the universal, macrocosmic secrets of the human mind, they hoped to uncover the specific, microcosmic secrets of their enemies.

These people have used their understanding of psychiatric methods to formulate and implement an action program based upon such beliefs. Once the neurotic map of each individual was determined, the government was able to set up a "filtering" mechanism, which are the different forms of brainwashing, to select various neurotic types and place them in their appropriate settings.

The psychopathic kernel of their long-range vision is:

> … converting the automized individual's world into a controlled environment.[18]

On one level, the technique is being applied to the world of intelligence, but on another level a far more hideous use has been envisioned by the practitioners of the art of brainwashing.

The most advanced among the brutal practitioners of this new industrial psychology was Dr. John Rawlings Rees, one of the founders of Tavistock Clinic, world's center for mass brainwashing and social engineering activities.

Rees discovered that an unreal realm could be created: the social group. An individual can be made to transfer his or her identity to the group, wherein they become subjected to the most intense forms of suggestion. Provided that the individual's inner sense of real identity is destroyed, he can then be manipulated like a child.

This is the objective of Social Networks. Welcome to the macabre world of Twitter.

Furthermore, what few people are aware of is that Twitter is now stored in the United States Library of Congress. I repeat, each tweet is transferred to the Library of Congress. You thought that your tweets are gone? No! Every tweet you've ever done is stored forever in the Library of Congress.

Nothing can be changed – what you put on the web, or tweet, or blog, or SMS (text), or a Web page or a Web page that you removed 15 seconds after the publication. Nothing. Once you have uploaded it, the system absorbs it into its own system, indexes it, and ties to your name – ready for someone to read it, to use it. This is data mining 2013, person-specific, instantaneous and ubiquitous.

What's more, every Tweet is constantly being monitored by marketing specialists, private detectives, intelligence agencies, governments, anti-terrorism experts, social scientists, Google, Microsoft, Amazon, etc. – whether your name is Santa Maria Goretti, Pope Francisco I or Ivan Ivanovitch Ivanov.

As of June 2013, Twitter has more than 500 million active users, sending 500 million tweets every day on every conceivable topic. The amount of information available through every one of your tweets boggles the mind, and with it, the profiling, the cataloging of you and every one of the people you are in touch with. It is done every minute of every day 24/7/365.

Most people think Twitter has only 140 characters when in fact every tweet is a goldmine of information. There's your IP-address, location, when you opened your Twitter account; in other words, 34-37 pieces of data in each Twitter tape.

Russia Today reported:

> … the US government requests for user data in the second half of 2012 is equivalent to just over 80 percent of all inquiries. Twenty percent of all US requests were 'under seal,' meaning that users were not notified that their information was accessed.[19]

Furthermore, Twitter interacts with Facebook. It's like a big invader of privacy to merge with a private life, and together they now mega-invade your privacy. But the fact that you're texting every hour on Twitter, updating your Facebook account, and constantly sending text messages, is not unusual. In fact, it is considered perfectly normal in today's society.

Twitter conversations are real-time statements from millions of people from moment to moment. They hold a wealth of information. How does the world's mood change throughout the day as a result of a particular event or a whole series of changes, orchestrated or otherwise? Twitter and social media allows a person to understand systems. It allows you to capture conversations on the societal scale, see them all and put them in perspective from a much higher plateau, something that was, up till now, impossible.

Group Think

Social marketing, used to be called "group think," the essence of which, is that you like what your friends like, and you vote the way your friends vote. *You are what your friends are.* That's why Facebook is worth so much money.

Facebook has more than one billion users, a seventh of the world's population. Facebook, with its clean image and reputation, gives the feeling that behind its logo you will find idealistic and young entrepreneurs, when in fact, those truly holding the reins of Facebook, are the elite.

In 2005, Mark Zuckerberg said,

> Facebook develops technologies that facilitate the sharing of information through the social graph; it is the digital zapping of the real-world social connections of users.

Wow!

The above mentioned social graph is nothing but pure thought control. Is that what Facebook is? Two members of the board of directors of Facebook are Peter Thiel and Jim Breyer. Thiel gave Zuckerberg $ 500,000 in 2004 to launch Facebook. Who is Peter Thiel? He is a member of the Steering Committee of Bilderebrg Group. Theil founded *Stanford Review* in 1987, and was a co-founder of PayPal in 1998.

Another Facebook board member, Jim Breyer, is also a board member of Anglo-American Accel Partners, a venture capital risk fund. Accel gave Zuckerberg $ 12.7 million in 2005 for development of a site where:

> … people can find other people like them, and find relevant information about their lives … their interests, information about your school, family, photos, likes and friends.

Now, why would a highly speculative, financial company be interested in controlling a social networking website?

Another character very involved with Facebook is Bill Gates. In September 2006, a month after signing the strategic partnership, Facebook introduced Minifeed and Newsfeed, two applications aimed at monitoring and reporting the real-time activity of each of the users of the network, even when the user is not connected to Facebook.

In November 2007, Gates invested $ 240 million to give twelve super-corporations access to the overall monitoring network. Among these twelve companies are Coca-Cola and video purveyor, Blockbuster.

How does this system work? For example: John Smith rents a video at a Blockbuster, and all information is sent immediately to Facebook and thereafter it becomes a part of Newsfeed. When any member of John Smith's circle of friends goes to his Facebook page they see immediately all their movements as well as those of John Smith and any other member of his network.

What's worse, this information is part of the instant composite sketch of John Smith, that super-corporations then store for later use in a limitless virtual repository. The goal is not only to get to know the user's profile, but also to know their way of thinking and acting. This is being done as we speak, all the time with every one of the billion users of Facebook.

However, Facebook isn't alone. Amazon offers their suggestions on books based on your profile, films you want to see, products you would want to buy for your loved ones. So does Facebook, MySpace, and countless other websites.

Mr. Thiel, in a speech at Stanford University in 2004, talked about accelerating change and bringing people to the virtual world ASAP. In Thiel´s unforgettable words:

> … controlling information is the same as controlling the human mind.

We may not realize it, but the elite certainly do.

Remember, everything that happens, everything you do, regardless of the importance of the event, is photographed, videoed, posted, discussed, tweeted, and voted on social media networks.

Then, there is MySpace. It absorbs a massive amount of information: name, date of birth, city, schools, places of work. Music - which, by the way, says a lot about you. What books you read – says a great deal about you. Your friends – tells people almost all they need to know about you. Where you live, your hobbies, your kids, your parents, your brothers and sisters, family ties, and photos, photos, photos are a cornerstone for investigators. Another cornerstone – your permanent location.

The big question as far as social media networks are concerned, "Who are your friends?" Again, if they know who your friends are, they know everything about you. And this is the very essence of Facebook.

Over one billion people use Facebook; that is 20% of all Internet users in the world. About 43% of Americans use Facebook. This is mind-boggling penetration. What is Facebook doing? According to *Wired* magazine: "The colonization of the Web."

What is the fundamental difference between Facebook and Google? Google indexes the entire Internet, and Facebook users indexes Facebook. And both of these corporations focus on each of us, individually, like a laser. They want to know – what you read, what you do, where you do it, who your friends are, your age, your gender, your religion, your orientation – everything. You are being indexed – catalogued

Facebook Connect and Open Stream are great business models that allow these mega-corporations to learn everything about you.

How many people have spent the time to read their license agreement with Facebook? Zero percent. Read it. You will be horrified to learn what you signed. Facebook simply swallows everything that somehow clarifies information about you. Facebook has bought FriendFeed; they interact with Twitter, and currently Facebook is buying travel sites because as soon as they find out where you are traveling, they want to be in a position to offer you tourist resources.

It goes without saying that every top Facebook executive has attended at least one Bilderberg meeting. The same can be said about top executives from Microsoft, Apple, and Google.

In 2012, Bilderberg discussed "How Do Sovereign States Collaborate in Cyber Space?" In 2011, one of the key items on the agenda dealt with, "Connectivity and the Diffusion of Power." The discussion specifically centered on the need to control the Internet as one of the key mechanism for overall control of society.

In 2010, Bilderberg vigorously discussed "Social Networking: From the Obama Campaign to the Iranian Revolution." The discussion focused on ways to use social media as agent of regime change throughout the world.

At the 2009 Bilderberg conference in Greece and at the 2008 conference in Chantilly, Virginia, the attendees examined "Cyber-terrorism: Strategy and Policy." The policy dealt with data mapping everyone – using the World Wide Web.

Then there is this about Microsoft. How many of you know what Cassandra is? Cassandra is a program that absorbs every drop of your information on Facebook. Microsoft needed to analyze and dissect you, so they created their own program – to invade of your privacy.

Have a look at the "Open Graph," connecting your Facebook account and your actions within Facebook with things outside of Facebook. Amazon contains your search history and the history of all your purchases. Which means your interests, your health and your political views are an open book. Never forget that every online action of yours falls into a database – and *never* disappears.

EBay, PayPal and Skype – how many people know that this is one company! EBay. This company knows everything about you. And most importantly, they know your financial information. They have your bank account, your credit card, and your home delivery address. This too is data mining, 2013 version – more powerful, more accurate and much faster than ever before.

The Scary World of Google

Let's start by saying that Google is not a lovey-dovey cute little company with a funny name, but rather a fierce, aggressive, predatory corporation. The term "Googol" is a word invented in 1938 by a 9-year-old Milton Sirotta, nephew to a famous American mathematician, Edward Kasner. Googol is an infinite number, a 1 followed by 100 zeroes. Infinity has no limits, but then, neither does Google or did you truly think that the owners of the company simply pulled a name out of a hat? Remember, limitless anything, spells c-o-n-t-r-o-l.

Corporation exists for the sake of profit. Google is also an integral part of the United States security apparatus. It collects and integrates everything that's already in the Google ecosystem: your messages, activities calendar, location data, search preferences, contacts, and personal habits based on Gmail chatter, and search queries. This information is processed, analyzed and stored for later use. Googol = infinity = control.

In January 2012, Google announced plans to integrate users' information across Gmail, YouTube, Google search and 57 other Google services such as Google Chat, GTalk, Google news, Google Maps, Google Music, Google finance, Google Checkout (PayPal competitor), and Google Video. That was one year ago.

Another of Google's new toys is their Google Goggles, an augmented reality. The product allows a user to take a picture of a picture, and it will tell you everything anyone needs to know about this picture. Everyone will want to know what your phone sees. So they will encourage the use of it: scan bar codes, people, places and the unilateral identification of different images (things and actions) – an augmented, annotated reality.

How can they do all of this? Because they own Picasa, an image viewer for organizing and editing digital photos. Combine that with face recognition technology, and Google, a Bilderberg-influenced company, has the technology to instantaneously recognize, identify, process and catalogue everyone who is in these photographs!

Again, how many people have read the Terms of Service for Google mobile? Zero! You, the consumer, give them the right to keep a permanent record, store, archive, and resell your location. And since GPS and Skyhook are now on most phones, they always know your location – within 3 meters!

Another revolutionary products is Google TV. How many of you know that Google is currently working with Logitech, Sony, and Dish network, to create a new series of applications? Why is this important? Because these are Bilderberg-affiliated corporations. Google will know when you sleep, when you wake up, when you watch pornography or whatever you see, while sitting in front of the TV.

What you watch, says a lot about you. How much time you spend watching TV, says a lot about you. When you get home, says a lot about you.

In March 2012, Google filed a petition for:

> Advertising based on environmental conditions ... using temperature, humidity, light and air composition....

In other words, Google is planning to use the ambient background noise of a person's environment to build a psychological profile of your entire life – opening a Pandora's Box of surveillance opportunities.

At the end of December 2012, a new service from Google:

> ... merged offline consumer info with online intelligence, allowing advertisers to target users based on what they do at the keyboard and at the mall. Without much fanfare, Google announced news of a new advertising project, Conversions API, that will let businesses build all-encompassing user profiles based off of not just what users search for on the Web, but what they purchase outside of the home.[20]

Why do they need to know so much about you? Control and power. The more they control the population, the more powerful they are. Google is no longer a company or even a mega-corporation. Google has become, for all intents and purposes, an all-seeing eye; what conspiracy theorists like to call: the New World Order. Not quite, but you can see it from there.

But Google isn't the only one. Verizon recently filed a patent for:

> ... gesture recognition technology.

This would mean that your TV would effectively spy on you 24/7.

> Verizon's technology would operate in the same way Google targets Gmail users based on the content of their emails – only transposing that principle into the home by scanning conversations of viewers that are within a 'detection zone' near their TV, including telephone conversations.[21]

Former CIA Director David Petraeus lauded this development as *transformational*, "because it would open up a world of new opportunities for 'clandestine tradecraft,' or in other words, make it easier for intelligence agencies and governments to spy on you."[22]

"Once upon a time, spies had to place a bug in your chandelier to hear your conversation. With the rise of the 'smart home,' you'd be sending tagged, geo-located data that a spy agency can intercept in real time when you use the lighting app on your phone to adjust your living room's ambiance," reports *Wired* magazine.

"Items of interest will be located, identified, monitored, and remotely controlled through technologies such as radio-frequency identification, sensor networks, tiny embedded servers, and energy harvesters – all connected to the next-generation Internet using abundant, low cost, and high-power computing – the latter now going to cloud computing, in many areas greater and greater supercomputing, and, ultimately, heading to quantum computing," Petraeus told attendees at a meeting for the CIA's non-profit venture capital firm In-Q-Tel.

Murky Relationships

In 2006, it was revealed that AT&T provided the National Security Agency (NSA) full access to its customers' Internet traffic. The data mining equipment was installed in a NSA backdoor, the NARUS STA6400, developed by Narus, a wholly owned subsidiary of Boeing, whose partners were funded by In-Q-Tel.

But as potentially as alarming as this all-pervasive monitoring is, and as disturbing as Google's interest in data mining technologies might be, the CIA venture capital arm is interested in more than just web-traffic monitoring and your holiday photos.

In-Q-Tel was formed by the CIA in 1999, with a mission to:

> … delivering technology to America's intelligence community. Publicly, In-Q-Tel markets itself as an innovative way to leverage the power of the private sector by identifying key emerging technologies and providing companies with the funding to bring those technologies to market. In reality, however, what In-Q-Tel represents is a dangerous blurring of the lines between the public and private sectors in a way that makes it difficult to tell where the American intelligence community ends and the IT sector begins.[23]

So, who is behind the CIA front company? Founding CEO of In-Q-Tel is Gilman Louie, member of the Markle Foundation Task Force on National Security in the Information Age – a CIA front company. According to Markle Foundation, their strategic mission is:

> … to develop revolutionary technology for the world's largest repository in order to create a virtual, centralized database.

Louie's long time business partner has been Jim Breyer, who shares the Board of Accel-KKR with the founders of Kohlberg, Kravis, and Roberts, (KKR), an equity firm who made a name for themselves in the 1980s for economy-destroying leveraged buyouts. KKR is one of the most important Bilderberg run corporations, whose senior partner, Henry Kravis annually attends Bilderberg meetings.

> Breyer and Louie also work closely with Defense Advanced Research Projects Agency (DARPA) board member Anita Jones, who sat on the board of In-Q-Tel with Louie. DARPA is not only leading the effort to create human cyborgs for Dick Cheney's perpetual wars, but also was the creator of the Information Awareness Office (IAO) in 2002. According to DARPA's own fact file, using 9/11, of course, as pretext: 'The most serious asymmetric threat facing the U.S. is terrorism. This threat is characterized by collections of people loosely organized in shadowy networks that are difficult to identify and define. These networks must be detected, identified, and tracked.'[24]

All of these challenges are clearly identified in the *Strategic Trends 2007-2036* report:

> The growing pervasiveness of ICT [Information Communications Technology] will require a concerted, comprehensive application of all the instruments and agencies of state power, together with cooperation from all relevant authorities and organizations involved in settling a crisis or resolving a conflict.[25]

So, what are these "concerted, comprehensive applications of all the instruments and agencies of state power?" The updated, scrubbed, renewed and made over version of Pentagon's Total Information Awareness, disguised as user-friendly applications.

For example:

> As a result of information extracted from wide-band monitoring, real-time tracking of cell phones is pretty much standard practice. Each call can be recorded verbatim and analyzed in real-time. As each number is learned the system fills in personal information. This can provide a

Google-Earth like view of cell phone movements, most of which can be remotely activated for espionage. Again, this web can be cross-reference with other webs of information.

Radios can reveal which frequency is being listened to. So, by plotting schedules of something like a Numbers Station, a satellite network can pinpoint a listener to within a few meters. Similarly, if you were to listen to a radio station sympathetic to a particular group, your location can be flagged and cross-referenced with information on current occupants.

Human beings emit radio waves in the ELF spectrum. The security services do not need to place a bug on us to track where we are, what we are talking about, what we are seeing or even what we are thinking. All of this information is being leaked into our environment 24/7 by our own bodies. All it takes is the right equipment to convert those signals into intelligence. It's no bigger a task than listening to a telephone exchange leaking radio waves and reconstructing the data into voice or data transmissions.[26]

Android phones make screenshots. Not you making screen shots of your private photos on your phone, but the Android itself takes the screen shots without your permission. Regular screenshots of everything you do on your phone is stored in the memory. Again, they don't ask for your permission, they simply do it. Are you are upset at this invasion of privacy? You shouldn't be. You gave them the right to spy on you. You don't believe me? Read the terms of service. They have the right to spy on you and you cannot do anything to stop them because you do not want to live without your smart phone, even if it has infringed on every one of your *unalienable* rights.

iPhone – Apple is a key corporation for the Bilderberg Group. Remember, all of these entities are not independently run, especially with Steve Jobs deceased. The Bilderberg-affiliated companies form part of a continuum; a dynamic system that changes with the times, absorbs and creates new parts, while excreting the remains of the decaying parts. Members come and go, but the system itself has not changed. It is a self-perpetuating system, a virtual spider web of interlocked financial, political, economic and industry interests based on the old Venetian ultramontane *fondi* model at the center. "Fondi" means pond in Italian, we are talking here about a system where shared financing leads to shared goals.

Venice today is the supranational homeland of a New Dark Age gang. Venice is a unifying symbol for the most extreme Utopian lunatic fringe and ideological fanaticism that radiates from numerous foundations, think tanks, private and semi private organizations acting as agents of a depopulation and deindustrialization agenda. The names change but the business model and the objectives always remain the same.

Apple and Google may seemingly compete for the same souls as clients, but in actuality they form part of a continuum that is closely collaborating with the elite on controlling the world. Through interlocked directorships and investments in common projects, these corporations work closely with powerful worldwide and US think thanks and foundations such as the: RAND Corporation, Hoover Institution, Hudson Institute, Brookings, American Enterprise Institute, Ford Foundation, Carnegie Endowment, Rockefeller Foundation, Aspen Institute, Club of Rome, Pilgrims Society, Atlantic Council, World Resources Institute, Council of Americas, Gorbachev Foundation, National Endowment for Democracy and literally thousands of others across the planet.

Apple, along with most companies, data mines every bit of information available to it. You are the focus. Your habits are the focus. You are the guinea pig. They know everything about you. Everything.

It seems less is known about Yahoo than about Google, but they have almost as many email accounts as Gmail. They have Yahoo groups, My Yahoo, web hosting, e-commerce, and are well positioned in Europe and Asia. Yahoo, immediately warns you – "When you register with Yahoo, and come into our services, you are not anonymous." Think about the meaning.

Your browser with its plug-ins, your actions, cookies, the trail you leave behind while you surf the Net, is absolutely unique. Your browser acts as an identifier, a digital identifier of you. But, there is more. Many things that identify you are located on the server and not on your computer. Put that in context, especially now that Google (Google Fiber) has become a full-fledged Internet provider.

Clue in, people, the truth is that all the communications happening on the planet right now can be monitored in near real-time.

Why is all of this being done: to control you, to dumb you down, to brainwash you, to influence you, to predict your future behavior – to turn you into a touchy-feely adult with infantile tendencies. *So you don't get in the way of important people by doing too much thinking on your own.* Think about it – if you still know how!

I am being serious.

Over 85% of people get all their information from the television and social media networks. In fact, the only "truth" most people know is what they get over the television or social media networks. There is now an entire generation of people who have not know anything that didn't come out of this tube.

This medium has become the Gospel, the ultimate revelation. It can make or break presidents and Prime Ministers. This tube is the most awesome force in this god-damn world. However, what would happen if it got into the hands of the wrong people? And when the largest company in the world controls the most awesome propaganda force in the entire Universe, who knows what shit will be peddled for truth?

Let's put it this way: "The advent and mass dissemination of television technology has rendered the Nazi model for a fascist society obsolete; it has provided a better, more subtle, and more powerful means of social control than the organized terror of the Nazi state"[27] and help subtly advance the cause of world government – without even explicitly stating that world government or One World Company is the goal.

Brainwashers call it "institutional human aggressiveness," what people like Freud say proves that people, us, are animals driven towards destruction. According to Freud, these aggressive destructive drives are "part of man's animal nature." The purpose of society, according to Freud, is to:

> ... regulate and control through various forms of coercion the outbursts of this innate bestiality against which the human mind is ultimately powerless.

Freud's principal point was that:

> ... masses of people can be organized around appeals to the emotions. The most powerful such appeals are to the unconscious, which has the power to dominate and throw aside reason.

Therefore, the key to mass brainwashing is to create an organized, controlled environment in which:

> ... stress and tension can be applied to break down morally informed judgment, thereby making an individual more susceptible to suggestion.[28]

And that was before the advent of Social Media, which renders all of the above mentioned systems obsolete. Remember: giant multinational media groups control not only television networks and your local newspapers and radio stations but also *every* major social media network in the world.

GLOBAL MEDIA MONOPOLY

AOL-Time Warner controls 292 separate companies and subsidiaries.

> Of these, twenty-two are joint ventures with other major corporations involved in varying degrees with media operations. These partners include 3Com, EBay, Hewlett-Packard, Citigroup, Ticketmaster, American Express, Homestore, Sony, Viva, Bertelsmann, Polygram, and Amazon.com. Some of the more familiar fully owned prop-

erties of Time Warner include Book-of-the-Month Club; Little, Brown publishers; Time, Life and People magazine as well as DC Comics. HBO, with its seven channels; CNN; seven specialized and foreign-language channels; Road Runner; Warner Brothers Studios as well as New Line and Fine Line Features in cinema entertainment. More than 40 music labels including Warner Bros, Atlantic and Elektra. Weight Watchers; Popular Science; and fifty-two different record labels.[29]

Viacom is controlled by Sumner Redstone, a perennial Bilderberg attendee. This conglomerate controls CBS, MTV, MTV2, UPN, VH1, Showtime, Nickelodeon, Comedy Central, TNN, CMT, BET, Paramount Pictures, Nickelodeon Movies, MTV Films and Blockbuster Videos, as well as 1800 screens in theaters through the Famous Players chain.

Disney owns eight movie-production studios and distributors: Walt Disney Pictures, Touchstone Pictures, Miramax, Buena Vista Home Video, Buena Vista Home Entertainment, Buena Vista International, Hollywood Pictures, and Caravan Pictures. The Walt Disney Company controls eight book house imprints under Walt Disney Company Book Publishing and ABC Publishing Group; seventeen magazines; the ABC Television Network, with ten owned and operated stations of its own including in the five top markets; thirty radio stations, including all the major markets; eleven cable channels, including Disney, ESPN (jointly), A&E, and the History Channel; thirteen international broadcast channels stretching from Australia to Brazil; seven production and sports units around the world; and seventeen Internet sites, including the ABC group, ESPN. sportszone, NFL.com, NBAZ.com, and NASCAR.com.

Vivendi Universal owns 27% of US music sales, labels include: Interscope, Geffen, A&M, Island, Def Jam, MCA, Mercury, Motown and Universal. Universal Studios, Studio Canal, Polygram Films, Canal +, numerous internet and cell phone companies, not to mention such artists as Lady Gaga, The Black Eyed Peas, Lil Wayne, Rihanna, Mariah Carey, Jay-Z.

Sony owns Columbia Pictures, Screen Gems, Sony Pictures Classics, 15% of US Music sales, labels include Columbia, Epic, Sony, Arista, Jive and RCA Records as artists Beyonce, Shakira, Michael Jackson, Alicia Keys, Christina Aguilera."

These international celebrities with their supposedly different viewpoints and ideas directly influence the general public. It also means that a single message – always presented from different angles – can easily saturate all forms of media to generate consent (i.e. "Arabs are terrorists.").

The Thomson Corporation, with its headquarters in Toronto, Canada, owns 105 daily and 26 weekly papers in the United States, mostly in smaller markets, not dominated by large city press.

The Pearson Group run from London is a $3 billion empire, and is one of the most powerful media influences in the British Empire and the world. It owns several papers, with its flagship being the City of London's most important journal, the *Financial Times*; it holds half ownership in *The Economist* magazine.

Fox News, part of News Corp, with their daily conservative rallies for couch potatoes, is owned by Rupert Murdoch, who owns a significant part of the world's media, amongst them his flagship publication, the *Wall Street Journal* and 20th Century Fox movie studio. Murdoch's Media Empire has been the main propaganda outlet for the perpetual war of the neo-cons and the Nazi minions.

Reuters News Media operates the largest newswire service in the world, with the world's most extensive international private satellite and cable communications network. Its news services are produced in 19 languages, with nearly every major news media outlet in the world taking one or more of its feeds. In addition, Reuters provides packaged coverage to hundreds of thousands of media outlets worldwide, providing features as well as news materials. Reuters Television is the world's largest international television news agency, reaching 500 million households, through 650 broadcasters in over 80 countries.

In the United States, the *New York Times* and the *Washington Post* are key media organs of the power elite linked to the Bilderberg Group. The *Washington Post*, founded by Eugene Meyer is the voice of official Washington, and has always stood for a weak presidency and ultimately a strong Federal Reserve – a private banking corporation.

The *New York Times* has the largest newsgathering force in the world and is twice as big as its nearest competitor. The *Times* also owns the *International Herald Tribune*, which is sold in 164 countries, with a daily circulation of several million. The *Times* has always been a Bilderberg and a British Crown propaganda channel, since Bilderberg's first meetings in the mid 1950s.

Historically, the *New York Times*, with utterly unwarranted self-assurance, designated itself the arbiter of 'All the news that's fit to print.' The *New York Times* has served the interests of the Rockefeller family in the context of a long-standing relationship. The current *New York Times* chairman Arthur Sulzberger Jr. is a member of the Council on Foreign Relations, the son of Arthur Ochs Sulzberger and grandson of Arthur Hays Sulzberger who served as a Trustee for the Rockefeller Foundation. Ethan Bronner, deputy foreign editor of the *New York Times* as well as columnist, Thomas Friedman among many others are members of both the Bilderberg Group and Council on Foreign Relations.

These media heavyweights do not merely analyze and interpret foreign policy for the United States; they help make it. Rather than offering an independent perspective on the actions of the politicians, the Establishment media acts as the ruling elite's voice – conditioning the public to accept, and even embrace, insider designs that otherwise might not be politically attainable.

Another media heavyweight is London's *The Economist*.

The Economist, mouthpiece for the City of London, started publishing in the heyday of the British East Indian Company in 1843. *The Economist*'s Editor-in-Chief, John Micklethwait was a participant in eight of the last ten Bilderberg conferences. Adrian Wooldridge, *The Economist*'s management editor is a habitual attendee at annual Bilderberg conferences. Nevertheless, nary a word is found in this prestigious establishment journal about the Bilderberg Group.

The Economist is jointly owned by Britain's Rothschild family and Lazard Frères banking houses, both close to Britain's royal family. Lazard is a leading French and British asset of the Anglo-Dutch combine, centered in the Royal Dutch Shell, which is the Dutch and the British Royal families and in the Rothschild banking organization. This is the phenomenon today called the Bilderberg Group.

Hyppolyte Worms, the founder of Banque Worms, itself a creation of Lazard, was a shipping magnate, whose business was built on its contracts to deliver Royal Dutch Shell oil.[30] He was also one of the twelve founding members of the Synarchist Movement of Empires, a secret organization behind the delivery of France to Hitler and the Nazis. Lazard Frères was the French investment bank for Shell, and it was in that capacity that Lazard was instrumental in the creation of the banking arm of the Worms group, *Banque Worms et Cie.*

Even in the age of Internet, blogospheres and other new generation contraptions of mass dissemination of information, the mass media still sets the tone for most news coverage, defining issues and setting the limits of "respectable" opinion. Journalism itself no longer exists, but rather promtional campaigns negotiated by PR firms running product advertising campaigns masked as news.

Associated Press (AP) is the oldest and largest news agency in the world. On any given day, it delivers some 20 million words and thousands of visuals, globally. It also provides a selectable stock service, an array of audio, text and information services. It also operates, via satellite, a nationwide radio news service for several hundred radio stations – making it one of the largest radio networks in the United States. AP's domestic network incluldes 143 bureaus and over 6,000 radio stations, and, through its international feed, to thousands more outlets overseas.

Then, there is Corporation for Public Broadcasting, PBS, supposedly is a public institution. "Available to 99 percent of

American homes with television, PBS serves nearly 90 million people each week."[31]

Every one of the aforementioned media groups and companies are members of the Bilderberg Group. You would never know it, however, as this information is never revealed to the public. The mass media serves as agents of change – the change being our inefficient understanding of the world around us.

The Fascist Concept of Man

The Nazi state was created by the same oligarchic financial and political interests who today control what we call the mass media and television. At the root of this experiment was the desire to create a New World Order based on reversing a fundamental premise of western Christian civilization: that man is created as a higher and distinct species from animals, created in the image of the living God and by Divine grace, imparted the Divine Spark of reason.[32]

What makes man human is our power of reason. The only thing greater than life: the power of the human mind. This is how mankind is measured. What separates us from animals is our ability to discover universal physical principles. It allows us to innovate, which subsequently improves the lives of people. Development of mankind, the development of the power of the individual and the nation depends upon scientific developments, upon seeking and discovering the Truth as our highest goal, thus perfecting our existence. Truth always lies in the higher order of processes. True sovereignty lies not in popular opinion, but in the creative powers of the individual human mind.

It's a moral problem, and a problem of destiny, every generation must advance beyond that of the preceding generation. And the hope, that it will happen, that is what should be on the mind of the person who is dying of old age: Has their life meant something, because it laid the foundation for a better life than they knew?

This is a fundamental clash of ideals. Those who measure up to a Renaissance view of a man versus those who see themselves, by birthright, above other men – who see mankind as an animal, whose worst impulses must be repressed by the state. This is a view from the Enlightenment movement, and in its extreme form, the fascist state. To have mass brainwashing work it *must attack* the Renaissance view of man, for no person seeking Truth, especially with a strong moral compass, can be brainwashed.

Endnotes

1   How to Profile the Terrorist, *EIR*, Volume 5, number 37, September 26, 1978.
2   Ibid.
3   Lonnie Wolfe, "Turn off your TV," *New Federalist*, p.5, 1997.
4   Edward Bernays, *Propaganda*, 1928, reprint, Ig Publshing, 2004.
5   Harley Schlanger, "Who owns your culture?," *Fidelio*, Vol. XII, No. I Summer 2003.
6   Ibid.
7   Lyndon LaRouche, Star Wars and Littleton, June 11, 1999, *EIR*.
8   John Chancellor with Walter R. Mears, *The New News Business*, New York: HarperPerennial, 1995.
9   Jeffrey Steinberg, "The Cartelization of the News Industry, "*The American Almanac*, May 5, 1997.
10   Andrew Gavin Marshall, "America's Strategic Repression of the 'Arab Awakening' Part 2," globalresearch.ca, February 9, 2011.
11   http://thesocialskinny.com/100-social-media-mobile-and-internet-statistics-for-2012/.
12   Ibid.
13   Bertrand Russell, *The Impact of Science on Society*, reprint edition, Routledge, 1985, ISBN: 978-0415109062.
14   http://thesocialskinny.com/99-new-social-media-stats-for-2012/.
15   http://www.skyhookwireless.com/location-technology/.
16   "The UN asks for control over the world´s internet," *RT*, December 5, 2012.
17   Manly P. Hall, *Lectures on Ancient Philosophy*.
18   "Rockefeller´s Fascism with a Democratic Face," *The Campaigner*, Vol. 8, #1-2, November-December 1974, p.56.
19   "Data Privacy Day 2013: Twitter reveals US government makes 80% of info requests," *RT News*, January 29, 2013.
20   Jason Lee, "Google starts watching what you do off the Internet too," Reuters, December 20, 2012.
21   "Verizon Files Patent for Creepy Device To Watch You While You Watch TV," Ryan Gallagher, *Slate* magazine, May 5, 2012.

22	Paul Joseph Watson, "You Read It Here First: Google's 'Ambient Background' Spy Tech, 'Internet of things' also a surveillance tool for authorities," PrisonPlanet.com, March 23, 2012.

23	Sibel Edmonds, "Google, Facebook, the IT Sector and the CIA," boilingfrogspost.com, October 5, 2011.

24	Nich Walsh, "Facebook: A Tombstone with a picture attached," *EIR*, December 7, 2007.

25	*DCDC Strategic Trends Report*, p.63.

26	"The NSA – Behind the Curtain," *Deep Thought*, January 16, 2012.

27	Lonnie Wolfe, "Turn off your TV," *New Federalist*, p.5, 1997.

28	Ibid, p.9, 1997.

29	Ben Bagdikan, *The New Media Monopoly*.

30	Gabrielle Peut, "The Howard Family: Stooges for the Synarchy," *The New Citizen*, July/August 2006.

31	www.pbs.org.

32	Lonnie Wolfe, "Turn off your TV," *New Federalist*, p.5, 1997.

Chapter Four

Space Exploration

We live in an infinite world. It seems like this doesn't have any real consequence; but let's imagine the opposite. Let's imagine the world is finite. Then we have to admit right away that the resources available are also finite, as well as the territory to be shared."[1] As human population grows, mankind is making an ever increasing impact on the environment – colonizing its land, consuming its resources, and releasing huge quantities of waste into its seas and atmosphere.

Now, let's extrapolate that finite existence into the future, to the very end, a generation or two from now. Imagine the wreckage and now, let's piece it together. What on Earth happened to us?

Ours was the greatest civilization in history, so advanced and powerful, it dwarfed anything that came before it, but like other great societies, it did not last. To have lived, and then to have died is not to be absent but to become absent; to be someone and then go away – leaving traces. How could the civilization that had mastered the planet, suddenly collapse?

We are back in the year 2015. Decay is rampant and unavoidable. Businesses are failing, and companies that remain in business face shortages and delays. People respond with a helpless sense of doom. The anxiety leaves one gasping for breath, and represents a declaration of defenselessness before a force too terrifying and massive to combat or even comprehend – a pervasive hopelessness and loss of spirit.

Great cities lay abandoned, incredible feats of engineering left to ruin. A collapse that caused the greatest disaster in human history – our own extinction.

If you are trying to assembly a multi-dimensional case, to understand what kind of a force drives events: a collapse of civilization, happening it seems with the precise stroke of a diamond cutter's knife – a pogrom of demand destruction designed to reduce the world's population in order to preserve for the elite the ever diminishing natural resources. This single piece of the puzzle was what started to make everything else resonate and make sense of the hidden dynamics

We know what a collapse looks like: Budapest's cobbled streets – a war zone. Protesters armed with blocks of ice, caught on film smashing up Hungary's finance ministry. Thousands trying to force their way into the legislature.

This is real. In the year 2015, the economic collapse is hitting hard in every industrialized country in the world. Around the world, emerging financial markets are imploding at a speed of light. The meltdown has hit turbo charge in Europe as a result of a three-week old lack of Russian natural gas.

Triggered by the economic collapse and compounded by human suffering in unheated, near-zero weather, riots have erupted from Latvia in the North, to Sofia in the South. Around the world, from China, to India, to Europe, industrialized nations are frantically preparing for civil unrest. This is not some piece of fiction. This is not *Atlas Shrugged*. This is about now. It affects all of humanity.

Another image: Ordinary people enraged by austerity cuts and draconian wage deflation, their hard earned savings reduced to nothing under forced government devaluation, fighting for their own survival. Civil unrest now moves from the back to the front burner. Political leaders and opposition groups from as far away as South Korea and Turkey, Philippines, Hungary, Germany, Austria, France, Mexico and Canada are calling for the dissolution of national parliaments.

This is madness, but it is real. It is all around us. We see it on daily basis on television, read about it in the press, and see it with our own eyes.

The European Monetary Union has left half of Europe trapped in a depression. Bond markets in the Mediterranean region are at all time lows. S&P has downgraded Greek debt to junk and the country's social fabric is unraveling as the pain begins.

The Spanish, Portuguese, Italian and Irish governments are balking at paying their short-term debt, putting at risk the solvency of the world's financial system. Cyprus is being bailed out, its citizens prohibited from withdrawing its savings in a desperate effort by the European Union to stave off collapse.

A great ring of EU states stretching from Eastern Europe down across Mare Nostrum to the Celtic fringe are either in a 1930's depression already or soon will be. Each state is a victim of ill-advised economic policies foisted upon them by elites committed to Europe's monetary project – either in the European Monetary Union or preparing to join – the states are trapped.

However, the economic aspect of it is just one area. This is as much about geography as it is about politics. A new order is being created where geography and money are proving to be the ultimate trump cards: geography becomes *the* governing economic decision-making factor.

Geography is giving us our first major political tectonic fault line. From the Baltic south, through Greece, into Turkey, then fanning out across the Middle East. This is the new frontier of soon-to-be flaming unrest.

A snake eating its own tail for nutrition. It is the way money works … for now.

Is there a chance, under present circumstances for governments to join forces and save the world from oblivion? A Snowball's chance in hell! The scope of the crisis, as we are discovering, is simply beyond one's comprehension. Our collective impotence to address an entirely new conceptual set of nation-threatening issues runs the risk of being seen for what it is – trying to fix the unfixable.

How much time do we have? Three months, six months, one year tops. Then what? Whatever is coming beyond that time frame is coming at us as we speak. We may not have the luxury of an organized retreat. The system is broken. And it cannot be corrected while an unprecedented "elite-controlled" economic collapse is smashing down every wall between humanity and the unthinkable.

Knives are coming out and points of no return are fast approaching. If our situation goes much further we will soon know whether the United States and the rest of the world lives or dies. Moreover, we will know whether a civilized society is an option or an untenable dream. If it is not, then the barbarians at the gates will come, and they will bring with them mighty appetites.

First stage: Systemic breakdown that will cripple the economy. The world economy comes to a screeching halt. No welfare checks, no Social Security, no healthcare benefits, no food stamps for the poor and no money to pay the millions of government employees.

Panics would, within a few days, drive prices significantly skyward. And as supplies no longer meet demand, the market will become paralyzed by prices too high for the wheels of commerce and even daily living. The trucks will no longer pull into food stores.

Hoarding and uncertainty would trigger outages, violence and chaos. Police and military will be able to maintain order for only a short time, if at all.

The damage that several days' shortage and outage will do, could soon wreak permanent damage that starts with companies and consumers not paying their bills and employees not going to work. This would be the second stage.

The poor will be the first to suffer and they will suffer the most. *They will be the first to die.* Death of hundreds of millions. That's the final stage. It is very hard and very painful to get one's mind to accept this reality but Mother Nature does not grant time outs.

The problem is, humanity has no Plan B and it is now also too late to come up with a Plan C or Plan D. *Progress is what brings light out of darkness, civilization out of disorder, prosperity out of poverty. All of these essentials are being challenged and threatened.*

Wars, famine, disease, droughts, social unrest, depleting natural resources. How often have we chased the dream of progress only to see that dream perverted? Technology offers us strength, strength enables dominance, and dominance paves the way for abuse.

Technological advancements are not the end of the world, but merely seeds for change, and change never comes without pain. It's in our nature to want to rise above our limits. Every time we have met an obstacle, we've used creativity and ingenuity to overcome it. And isn't achieving a dream, worth it?

Society needs laws and regulations to protect it. And if the elite need to work in the shadows, pulling strings to enable us to head in a safe direction, would supporting them be all that bad?

We are at a zero hour, *facing our own demise as humanity*. What does death look like from the other side? People die at least twice. Once physically, once notionally; when the heart stops and when forgetting begins. The lucky ones, the great ones, are those whose second death is decently, perhaps indefinitely postponed.

Does death reveal that there has been no life, only a dream of life. And will the *new Man*, the one that comes after our society disappeared, remember us as we were, or think of us as victims of happenstance?

Time and space – the pile of debris we call history; also represents our successes. And we have had successes. Will they be sustained or will they vanish like time?

A thin pearl of light from the fading moon plunges into the angry swells of the dark, furious ocean; the white sprays are caught in a night sky rolled over the chiseled rock under the force of the night wind. And we are back in 2015 – at zero hour. Depleting natural resources and the ever-expanding population base has made it mission urgent to discover alternative energy sources.

Behind closed doors, in the shadows, the elite and their concubines are scheming, conspiring and working on secret survival plans for their own future. Natural resources are the corner stone of their plans. The Earth has been depleted, raped and devoured by the greed of men. Now, what if the greatest discovery of natural resources didn't take place on Earth? And if not on Earth, then where?

We have nearly devoured Planet Earth. The next challenge, before it is too late, is the heavens and space exploration. Who will get there first? We, the people or they, the world's elite?

> The extraterrestrial imperative is the next step of evolution. We are faced with possible extinction, either through thermonuclear war, or a dark age, because of the disintegration of the financial system. The question is, can we change and choose the alternative such as space colonization?[2]

With the world's population base exploding, our best hope for survival is an alternative energy source, possibly found in outer space. Without it, we are doomed.

> Krafft Ehricke, who in the Apollo program, was responsible for the development of the Centaur rocket and the Atlas rocket already, many decades ago, defined the industrialization of the Moon and the colonization of Mars as the goal because that subsumes virtually all necessary breakthroughs in the realm of science and culture which we need to master if we are to have a continued existence of civilization.[3]

First stop, Earth's satellite, the Moon. The research group at the Fusion Technology Institute at the University of Wisconsin are working out the details of a plan to begin lunar mining of helium-3:

> Not only for use in setting up lunar industries and powering Earth reactors, but also for advanced fusion propulsion systems to open the rest of the Solar System to man's exploration.[4]

Is this a Utopian dream? How will mankind get to the Moon and then to Mars? Well, we're not going to do it all at once. To get to Mars from Earth's orbit, we first need to get to the Moon. Once we get to the Moon, the first thing we need to do, is:

> Build a manufacturing facility on the Moon, which utilizes the raw materials on the Moon itself to develop the elements of materials and devices that you can ship to further destinations, such as Mars.[5]

And once built:

> Getting even massive components off the Moon's surface is far easier than it would be from Earth due to lower gravity and lack of air (it took a tremendous Saturn V rocket full of fuel to get to the Moon, but only the tiny Apollo

ascent module to get back off). Building vehicles and other space-based structures on the Moon is vastly easier and less expensive than it would be here on Earth. From there, the rest of the solar system is an easy trip.[6]

For example, the Moon harbors enormous resources that we can use on Earth, including titan, aluminum, and iron. Water from asteroids can fuel an in-space economy.

Dr. Gerald Kulcinski told a lunar science conference in September 1986:

> ... the Moon has a decisive advantage relative to the Earth, in the purification of those metals, which are always found in raw minerals that contain a lot of oxygen. On Earth, the molten metal must be placed in a vacuum to achieve oxygen extraction, thereby obtaining the best mechanical and anti-corrosive qualities possible. But to create that vacuum is very costly.
>
> Because the Moon has no atmosphere, the vacuum is free, and of a much better quality than anything we have been able to create on Earth. With a perfectly purified lunar titan, we could build bridges on Earth that would last forever. All this is possible only if the metal purification is achieved on the Moon.
>
> Furthermore, the Moon harbors important reserves of Helium-3, very rare on Earth, which is the ideal element to realize nuclear fusion ...[7]

The very same source of energy which is abundant in much of our Solar System and on the Moon. In fact:

> There is enough fuel on the Moon to meet the present energy needs of the entire Earth for close to 2,000 years ...

Helium-3 is a natural decay product of radioactive tritium and is the most effective, most efficient for the production of thermonuclear weapons.

> And thermonuclear fusion power is several orders of magnitude more powerful than any nuclear power. Therefore, it means we're making a leap in the amount of power avail-

able, per capita and per square kilometer, for the territory in the Earth, in the Moon, and so forth.[8]

Industrial materials processing on the Moon will be significantly different than conventional Earth technologies, which require vast amounts of water, chemicals, and other volatiles that do not exist in the lunar environment. Fusion has great advantages, even over nuclear fission, for materials processing and other industries: It requires a small amount of fuel, most of the fuel is already on the Moon, it produces no waste that requires recycling, it requires virtually no radiation protection, and it can make use of direct conversion technologies such as magnetohydrodynamics (MHD) – getting rid of the steam turbine. In fact, direct plasma processing, using the high-temperature charged-particle product of the fusion reaction itself, has the potential to increase productivity by orders of magnitude over today's chemical or even electrical processing technologies.

From a very small amount of matter can thus come an inconceivably abundant and non-polluting source of energy. Numerous other applications can be conceived on the Moon, such as the utilization of the phenomenon of free superconductivity, available due to the cold conditions that reign on Earth's satellite.[9]

Can the Earth be saved or have the elite given up on our planet while secretly planning a mass exodus from Earth? Are you surprised at the question? You shouldn't be.

As we speak, plans are being drawn up by the European Space Agency and Russia for a joint mission to Mars. Privately funded corporations such as Planetary Resources and Golden Spike Company are working around the clock on a landing mission to the Moon and the mission-critical projects of asteroid mining.

A private space venture from the Netherlands called "Mars One," aims to send four astronauts on a one-way journey to Mars in an experimental colony by the year 2023.

The British government's top secret *Strategic Trends Report* makes the intentions of the elite crystal clear. By the year 2050:

We are going to power ships that go in significant numbers with people in them, to Mars.

That's right, the elite are planning, at least, a limited exodus from the Planet Earth. Why? What do they know that we don't? Nuclear wars? Nanowars? Bacteriological wars? They would know. They are organized to operate over and above society.

What is known, according to a 20-year-old declassified NASA report, is the following.

Phase 1: Helium-3 Mining

Between now and the year 2020, a manned space station will have small vehicles parked outside that can go back and forth to the Moon. Their purpose? Helium-3 mining. Where will the helium-3 come from for their fusion reactors?

> The solar wind, which constantly spews high-energy particles and radiation from the Sun throughout the Solar System, has been found by spacecraft probes to contain about 20 parts per million particles of helium-3. However, these particles do not survive the Earth's atmosphere and are, therefore, not found deposited on the Earth's surface.
>
> Because the Moon has no atmosphere, the helium-3 bombarding it from the solar wind has collected there over billions of years. Samples of lunar soil returned by the Apollo astronauts and analyses from the Soviet unmanned Luna probes indicate that the lunar soil contains an estimated one million tons of helium-3.
>
> The helium on the Moon will not be an unlimited energy supply in itself, but it could function as a one-century bridge to the recovery of virtually limitless helium-3 from the outer planets. It will open the next millennia, providing humanity with the first biologically benign, non-polluting, efficient, and economical energy in human history. The abundance of this quality of energy will actually create the possibility. And is itself prerequisite for the colonization of space, and the necessary revolution of all economic activity on the Earth. The Moon can open the fusion era.[10]

Now, theoretically, with helium-3 as a fuel, you are approaching the possibility of a rate of acceleration – acceleration *of* acceleration – of an impulse toward Mars, which scientists have estimated at about three days, from Moon orbit to Mars. Without this mode of power, we can send things to Mars now, if we're willing to wait 300 days or longer for the arrival of that package from the Moon to Mars. But if you want to send a person there, 300 days in a spacecraft between here and Mars, is not recommended for the health of anybody. Their bones would sort of disappear, and if they were alive at all, they might end up there as a blob, and they probably would have some difficulty in making the return trip, if it were possible! But these problems are solvable.

So therefore, with the access to a thermonuclear fusion approach to the power base of action in the universe, there is no visible limit to what mankind might be able to do.[11]

But, there is another aspect of space exploration worth examining. A new paradigm for resource discovery. Private sector human expeditions to the Moon are now feasible, primarily using existing space systems or those in development, trying to bring the solar system into humanity's sphere of influence.

The Golden Spike Company is working to implement and operate a human space transportation system at commercially successful price points.[12]

Circling the Sun between Mars and Jupiter are near-limitless numbers of asteroids, chunks of rock ranging in size from footballs to gigantic Ceres, 1,000 km (620 miles) across.

More than 1,500 are as easy to reach as the Moon and are in similar orbits as Earth. Asteroids are filled with precious resources, everything from water to platinum.[13]

Many of the asteroids are on orbits that also take them near the Earth, which makes them far easier targets for space missions, requiring less fuel and time to visit.[14]

Water is the essence of life and exists in plentiful supply on asteroids. Access to water and other life-supporting volatiles in space provides hydration, breathable air, radiation shielding and even manufacturing capabilities. Water's elements, hydrogen and oxygen, can also be used to formulate rocket fuel. Using the resources of space – to explore space – will enable the large-scale space exploration.

Harnessing valuable minerals from a practically infinite source will provide stability on Earth, increase humanity's prosperity, and help establish and maintain human presence in space.[15]

Under such a project, NASA would use its big new rocket to get astronauts to the Earth-moon Lagrange-two point, where gravitational forces from both bodies cancel out and allow a spaceship to sit tight without expending fuel. From here, a crew could stay in continuous contact with mission control on Earth while floating 40,000 miles above the far side of the moon, an area never explored by Apollo. Perhaps as early as next decade, three astronauts could visit L2 in NASA's Orion spacecraft. There, they would meet up with a deep-space habitat derived from leftover ISS parts that NASA is currently planning.[16]

The goal of NASA is to conduct research to enable safe and productive human habitation of space as well as to use the space environment as a laboratory to test the fundamental principles of biology, physics, and chemistry. In the longer term, humans will venture beyond low earth orbit, first to explore Mars, following a path blazed by robotic systems.[17]

Technology and Space

We've already seen the kind of improvement that comes from this exponential growth of information technology. The Curiosity mission to Mars has incorporated into its roving vehicle: robots, lasers, chemical products, the best technology there is.

Communication technologies are trillions of times more powerful than they were a century ago, when we were sending Morse code over AM radio frequencies, for example. If you follow the inevitable trajectory of ongoing progress at an exponential pace, we get to a point where we may merge with this technology and greatly expand our own intelligence.

If we can apply these technological developments on Mars, we could also apply them to the civilian economy on Earth. And as we have seen with the Apollo project, the technological revolution related to space missions produce great benefits: Teflon, computer chips, robotic self-driving cars that build instant three-dimensional LIDAR scans of their surroundings and telerobotics.

For example, modern surgeons can now operate on a patient from across a room and officers in the Pentagon order a drone kill from across the world.

Technological development for mankind is a very big part of space exploration. Technologies are invented to make our lives easier, not our choices. The problem comes, when we forget that.

> The main way that telerobotics differs from current robotic exploration, such as that done by Curiosity and other rovers, is that it places human cognition right where the action is. Mocup, a diminutive Lego robot, is the first phase of an ambitious European Space Agency project called METERON, the Multi-Purpose End-To-End Robotic Operation Network.
>
> This project will test sophisticated communications and human-machine interface technology for future exploration. By 2014, ESA scientists hope to fly an exoskeleton to the International Space Station, and an astronaut would wear it over their arms to provide haptic feedback – essentially a sense of touch – to a robot on the ground.
>
> Such a skeleton could one day help astronauts with detailed construction projects, plugging in or connecting components, on the moon or Mars. This version of space exploration looks much more like the movie *Avatar* than *Star Trek*. In addition to saving money, it keeps humans out of harm's way.[18]

Pentagon's science division, DARPA agrees. For nearly 50 years, the Defense Advanced Research Projects Agency:

> … has engineered technological breakthroughs from the Internet to stealth jets. But in the early 1990s, as military strategists started worrying about how to defend against germ weapons, the agency began to get interested in biology.[19]

The agency is already researching ways for its troops to use their minds to remotely control androids that will take human soldiers' place on the battlefield. These are the direct derivatives of space age technology. According to the DARPA's 2013 budget:

> The Avatar program will develop interfaces and algorithms to enable a soldier to effectively partner with a semi-autonomous bi-pedal machine and allow it to act as the soldier's surrogate.
>
> The rise of the machines is here. We are facing the time when an unconscious evolution period is almost finished. And we come to the new period when a controlled human evolution can and will happen. Technological progress will be concentrated on making a new body for the human being.
>
> The plan is to incrementally move the human mind into more disembodied and futuristic vehicles: first a humanoid robot controlled entirely by a human brain via brain-machine interface, then a conscious human brain transplanted into a humanoid robot, then consciousness uploaded to a computer, and finally a hologram that contains a full conscious human mind.[20]

Super Soldier

DARPA is funding dozens of human augmentation projects around the world, to radically improve the performance, mental capacity, and resilience of soldiers. Everything from tapping the potential of their unconscious minds to augmented cognition and the creation of uninhabited combat vehicles whereby the military expects in the very near future for individual pilots to fly entire squadrons of robotic planes. All of these projects are a direct result of space age exploration technologies.

A controller could monitor a pilot's brain with functional near infrared spectroscopy. Another project being funded concerns less sleep. In this project, psychologists are using transcranial magnetic stimulation to counter fatigue. Other projects include stronger exoskeletons and tougher bodies.

> All sort of new techniques from pharmaceuticals to robotics exoskeletons make this vision come true. It is import-

ant to realize that this report clearly states that cybernetic enhancement of human performance is inevitable.[21]

For instance:

> ... the HULC is an anthropomorphic exoskeleton that mimics the human form. It provides extra support, enables the person to carry extra weight than he normally could. Exoskeleton technologies are part of robotics field. Think of it as wearable robots. They sense what the user wants to do, where the user wants to go and then makes the motion. The exoskeleton is primarily composed of titanium components, which are both lightweight and provide the strength needed to carry the load. A microprocessor takes the reading that is throughout the structure and calculates where the user wants to move and then commands the actual hydraulics system to activate the joints systems to provide the motion.
>
> This technology is a real benefit to the military. Just imagine your solder operating at 3000 meters in the Afghan mountains and being asked to take an 80kg bag up in that thin air. How exhausted you would be once you got there. An exoskeleton provides the ability to carry that weight the same distance but to have the energy left to execute the mission.
>
> The best thing about working on this technology is that it gives the army a glimpse into the art of the possible.[22]

Where is this leading the new generation of armed forces? What will they become? Will they be more than human or less? There is technology, there is strength. The temptation to misuse it is hard to resist. Will our soldiers be strong enough to stay human? Silence. Deathly silence.

Moth Cyborg

However, DARPA's vanguard projects are not based only on their research into super soldiers. The agency is working hard to establish an array of cybernetic insects. The Defense Advanced Research Projects Agency is implanting computer chips in moths while still in the pupa stage for use in surveillance. The moth grows around the chip and then its nervous system may be controlled by a remote control.

The program, the 'Hybrid Insect Micro-Electro-Mechanical Systems' (HI-MEMS) is part of their Controlled Biological and Biomimetic Systems program and seeks to:

> … fuse mechanical micro-technologies to living insects creating a machine-bug amalgam.[23]

It also includes outfitting other insects with miniscule sensors, cameras and a wireless transmitter *attached to their very nervous system that can be controlled remotely* like RC airplanes and could send data from places inaccessible to humans.

> The microsystems that DARPA calls payloads will draw parasitic power from the insects, work with the insect's nerve or muscles, and take control of the insect.[24]

As Fox News reports,

> It is hoped that one day, a sensor-enabled insect with a 100-yard range could be placed within five meters of a target using electronic remote control and, potentially, Global Positioning System technologies. Ultimately, the moth will be able to land in enemy camps in remote location unobserved, beaming video and other information back via what its developers refer to as a 'reliable tissue-machine interface.'
>
> This latest development will allow the moth cyborgs to spy on enemy insurgents, and is the most advanced robotic technology ever conceived by DARPA.[25]

The remote-controlled moths are just part of its overall research into microelectromechanical systems, were one of a number of technologies soon to be deployed in combat zones.

> The Department of Defense has said it wants one-third of all missions to be unmanned by 2015, and there's no doubt their things will become weaponized, so the question comes: Should they be given targeting authority?
>
> Biological engineering is coming. There are already more than 100,000 people with cochlear implants, which have a direct neural connection, and chips are being inserted in people's retinas to combat macular degeneration.[26]

Another iRobot project being developed as part of the US military's 'Future Combat Systems' program:

> …is a small, unmanned vehicle known as a SUGV, basically the next generation of the PackBot, one which could be dispatched in front of troops to gauge the threat in an urban environment. The 30-pound device, which can survive a drop of 30 feet onto concrete, has a small head with infra-red and regular cameras which send information back to a command unit, as well as an audio-sensing feature called 'Red Owl' which can determine the direction from which enemy fire originates.[27]

Let's face it. We are witnessing an unparalleled explosion in scientific knowledge.

> Because of the synergy, the interplay of three great revolutions: quantum, computer and biotech revolution, we have learned more in the past 50 years than in all of human history.
>
> The marvels of science like the Internet, telecommunication satellites, laser beams, radio, television, microwaves, even a structure of a DNA molecule, and biotechnology. All of it, ultimately, comes from the quantum theory.
>
> One of the byproducts of quantum revolution is teleportation. The first results of teleportation were achieved 10 years ago in Austria by an Institute for Experimental Physics and its lead physicist, professor Anton Zeilinger, when they teleported individual particles of light from one prism to another. It is called quantum entanglement. It is a mysterious bond between a pair of photons. When you change one, the other one changes instantly. In fact, the photon itself isn't teleported, just the information it contains. This information is teleported over to another system, which assumes that information, therefore becoming identical with the original. The original loses it properties and the new one gains it. What is amazing is that the original photons' information is lost, which could have huge implications for teleporting anything on a larger scale.
>
> Another excellent example of this synergy is the human genome project. We now use laser beams to scan hundreds of genes at any one given time. And then we

use super computers to process this vast amount of genetic information.

We have the means through genetics, robotics, information technology and nano technology to control matter and energy and life itself. We have never seen anything like this before and it is raising profound questions about what it means to be human. The intense cross pollination between the computer, biotech and quantum revolutions will give us unprecedented power in the twent-first century.[28]

The undisputed leader of this change is nano revolution.

Nanotechnology

Nanotechnology is a powerful new technology for taking apart and reconstructing nature at the atomic and molecular level. Think of computers, chemistry and material science. Computers have brought us to nano-electronics. Billions of nano components on a scale of 10s of nano meters. Material science has brought nano-scale structures. Chemistry has led to increasingly advanced and atomically precise molecular structures.

"Nano" is space measured between one atom and about 400 atoms. It is a space in which quantum behavior begins to replace the Newtonian physics. There is enough room for billions of small molecules. Molecules are well defined – precisely defined chemical entities. Nano particles are almost always different.

They have the gradations in size. Their surfaces are different in detailed textures, their compositions vary. They are not distinct entities, but a part of a continuum. Furthermore, the surfaces are modified by environmental exposure.

In the body, proteins stick to them. In nature, they can be chemically modified in various ways.

What are possibilities in terms of physical laws? In any direction, if you go far enough, you hit limits set by physical law. Quantum mechanics, speed of light, gravitational forces, etc.

Now, think of the Earth on a new scale, a billionth of a meter. The distance between the moon and the earth is on the order of a billion meters, a day's travel. The distance between a meter and a billionth of a meter, is roughly the same gulf, but takes a few seconds.

We are heading deep into the world of nano-science, down to the dimension of an atom. To put in context: a strand of DNA is 2.5nm wide; a protein molecule is 5nm, a red blood cell 7,000 nm.

Compared to the human hair, a nanometer is 100,000 times smaller than your strand of hair. At a nano-meter scale, everyday materials start to act in unimaginable ways. The behavior of nano materials can change when the size becomes so small as compared with a larger amount of that same material.

As these technologies come together, what we will be seeing is atomically precise nano-systems, small scale by microscopic standards, but large scale by the previous standards of atomically precise systems.

These nano-technologies have wide industrial use today. In medicine, they are being used to put functional structures into the body, whereby part of it targets the cancer cell and part of it contributes to destroying it.

> Nanotechnology embodies the dream that scientists can remake the world from the atom up, using atomic level manipulation to transform and construct a wide range of new materials, devices, living organisms and technological systems.[29]

The economic, military and everyday consequences would be immense. First of all, we can see increasing miniaturization of components in the electronic or nanoelectronic industry. The storing in tiny volumes of space far more processing power than is currently possible, but also making use of biomedical applications: an artificial retina, replacing an ear that no longer works, being able to make molecular wires, like a nerve which could create all sorts of impulses that the brain would decode.

With the advent of new science and technology related to the genetic revolution, we can literally rewrite our genetic construct. In a span of a generation, we have gone from genetically modified plants, to genetically modified animals. Is the next step genetically modified humans?

In fact, genetic engineering and nanotechnological military applications are the next arms race.

Metamaterials

Our mastery over properties of matter is massively opening up our horizons. For example:

> … nano structured photonic materials represent a paradigm shift in technology, and the beginning of a new photonic revolution. The industrial revolution changed the fabric of society, the optical light. An optical fibers carrying light created global information networks. Photonic metamaterials will bring the next revolution, providing smarter solutions for all industries where light is used: from telecommunication and data storage to defense security.[30]

In 2010, scientists created a material that should not exist in nature. When light hits an object, it is the object's atomic structure that determines what we see.

Whether it is translucent marble, clear water or green leaves. It all depends on how the light interacts with the atoms. If we can manipulate those atoms, we could ultimately control what the world looks like. How? By creating artificial materials called metamaterials.

> And these new structures outperform nature in quite startling ways. And they do so at incredibly small scales. Nanophotonic metamaterials provide a degree of control over light that was unimaginable even a few years ago.[31]

The first applications are obviously for military stealth. One direct derivative of the metamaterial that we couldn't imagine, just a few years ago, is an invisibility cloak. It doesn't exist in nature.

The retro-reflective-projection technology for an invisibility cloak was developed by a Japanese team at Keio University in 2003.

> What makes the technology unique is a fabric made of glass beads only 50 microns wide which can reflect light directly back to the source much like the movie screen. The material can be applied to almost anything. The goal is to create augmented reality, which allows anyone to easily see information on real world objects.[32]

In military applications, this invisibility cloak,[33] a sheet of hexagonal pixels called AdaptiV, helps military vehicles blend into their surroundings and avoid heat-seeking missiles because the metamaterial deflects the light in a way that light flows around a hidden object with no discernible distortion.

> The light you and I see with our eyes is just a small range of frequencies on a much larger spectrum called the electromagnetic spectrum. For example, glass is clear to us because it allows light in the visible frequency range to pass straight through it. If we move a little higher in the spectrum to ultraviolet and glass is no longer see through. X-rays are another example of this. In the X-ray frequency, our bodies are clear like glass, while our bones are not clear because they have a different structure than our skin and organs. When light interacts with matter a couple of different things can happen: you can have reflection, where light bounces off completely. You can have refraction, were light enters the material and passes through it or you can have a combination of the two.[34]

Then we have carbon nanotubes, made of material that is harder and lighter than steel.

> Carbon nanotubes are a miracle of nature. They are made up of individual carbon atoms arranged in a hollow cylinder. The cylinder's surface is just one atom across. The diameter is 50 atoms across and these tubes can be billions of atoms long. Their atoms are bonded with the strength of diamonds, yet they have the flexibility of fiber. In the future, we will be able to use carbon nanotubes for unsmashable card, uncollapsable building, ultra light jet planes as well as a highway into space.[35]

As Katherine Bourzac reports in *MIT Technology Review*:

> If an airplane painted with a nanotube coating were hit with a radar beam, nothing at all would bounce back, and it would appear as if nothing was there. Nano paint would make airplane invisible to radar. The heat from the sheet of nanotubes affects the optical properties of the surrounding

water, creating an illusion of invisibility. That means that, unlike other metamaterials that are more spectrum specific, they can cloak an object not only from visible light, but from things like radars as well. This is much cheaper than buying a stealth airplane and would produce a cost effective means for entire fleets.[36]

*****

Wars, any wars, big or small, are a great opportunity for money making on one hand and population reduction on the other. You sell guns that kill people, and killing people, you reduce their numbers. Population reduction. This is a win-win proposition as far as the elite are concerned.

An increasing number of people in the world mean a greater potential to cause trouble, and to demand a bigger slice of the "Big Pie of Life." The authors of the *Strategic Trends Report 2007-2036* devote a lot of space to Global Pandemics. According to the report:

> A new untreatable virus that spreads rapidly among human populations, causing serious illness, would pose a significant threat to societies, which are both increasingly concentrated in urban settlements and also connected by modern mass-transit systems. Such an illness would be extremely difficult to contain and could have catastrophic impacts beyond its immediate medical effects, possibly including the collapse of the highly integrated global economy, with its complex networks of interdependent relationships. Similar effects might be induced by the rapid spread of disease among livestock or crops.[37]

Too far-fetched? For example, In-Q-Tel, CIA's non-profit venture capital firm is spending some of its money on two areas of biotechnology: Nano-biology convergence and physiological intelligence. The company is interested in:

> … human genetic and biochemical response to disease or environmental exposure to chemicals or organisms.

Another company interested in human genetic and biochemical response is BioRad Laboratories in California. They make one of the most hazardous biological and nuclear chemical compounds known to man. It would have been Hitler's wet dream. The chemical compound is selective to the degree, in that with an appropriate genetic material, you could wipe out whole segments of humanity. It is race specific.

In-Q-Tel and BioRad Laboratories are both closely related to the US intelligence apparatus. Does anyone in his or her right mind believe this to be a coincidence? And these viruses, not necessarily need to be simple bio-viruses. They could be nano-viruses, which are a great deal more deadly.

Nanointelligence and Nanoweapons

Nanobots were created to be like life. To be able to reproduce and to serve our needs. The intelligence of nanotech will not be in one nanobot. It will be a collective intelligence of trillions of nanobots working together and pooling their thinking resources.

Dr. John Alexander, adviser to US special operations, said,

> You can program nanobots specifically so that they attack only people with certain characteristics.

They can be sent out to take out people in the most secure locations. Once programmed, nothing would stop them. But can we control them? The fear is that to breed, they will need organic material – which is everywhere. This means, they may reproduce themselves. And as their numbers increase, they can turn on the environment around them – as they begin to take on the life of their own.

These nano weapons attack a target at the molecular level, and then use the molecules of their target as raw materials to replicate more copies of themselves. In this way, small amounts of these replicators could have the capacity to destroy large targets very quickly. They could devour plants, animals, and ultimately, people. This is how wars may be fought in the future. Can this be our future? A world transformed … by good or evil?

Do you remember what I said at the beginning of the chapter? Ours was the greatest civilization in history, so advanced and powerful, it dwarfed anything that came before it, but like other great societies, it did not last. Great cities lay abandoned, incredible feats of engineering left to ruin. A collapse that caused the greatest disaster in human history – our own extinction. Could nanotechnology be harnessed to do this?

NANOTECH WEAPONRY – MODUS OPERANDI

Molecular manufacturing raises the possibility of horrifically effective weapons. As an example, the smallest insect is about 200 microns; this creates a plausible size estimate for a nanotechnology-built antipersonnel weapon capable of seeking and injecting toxin into unprotected humans. The human lethal dose of botulism toxin is about 100 nanograms, or about 1/100 the volume of the weapon. As many as 50 billion toxin-carrying devices—theoretically enough to kill every human on earth—could be packed into a single suitcase.[38]

These nanobots are small enough to inhale, small enough to enter your body through your skin, small enough to circulate in the body, cross the blood brain barrier,

> … spread through food and water, and be designed to kill certain ethnicities. That is just about the worst thing that nanobots can be programmed to do. If nanotechnology is so far along that nanobots are able to fly and navigate the human body then it is only logical that nanobots can also fly and watch out for hostile bots. Keep that in mind when it seems like the targets of these weapons are defenseless.[39]

In a well-written piece of nanotechnological weapons, Kevin Robert Baker reports, "a nanobot could be dropped into someone's food which is designed to hurt the host. It only takes 100 nanograms of Botulism to kill a human. The nanobot can wait until it has reached the brain, and then deploy its payload. The results would be flawless and devastating.

> In stealth operations the nanobot could induce strokes or any other body failures. A nanobot that destroys an artery

in the brain can be confused as a brain aneurysm. Don't forget the possibilities of nanobombs. Those won't go very well with your brain or lungs.

A nanobot can fly into an enemies eye and burrow itself until it reaches a critical part such as the optical nerve. Once there, the nanobot can damage the nerve. Although the enemy would be blinded temporarily; nanotechnology would be advanced enough to be able to repair the nerve. These nanobots could also destroy any nerve in the body such as the spinal nerves. Nanobots could also be used to destroy pain receptors. Something that will make soldiers feel invincible.

Imagine small nanobot deployed with one purpose. To infiltrate a computer and monitor the frequencies picked up in the RAM. It may also steal the information off of the hard drive. The Pentagon already worries that devices can be built to monitor your computer screen through only the frequencies given off by the display.

How could this be done? This particular nanobot could drop off signal transmitters as it travels towards the computer to be monitored. Once there it sends out the frequencies to the other nanobots that eventually reach its controller. There the information could be reconstructed to show the sensitive data. Although computers can be easily hacked, nanobots would provide access to networks that are not connected to the Internet.

Nanocameras could also be attached to an infinite amount of flies and animals which could spy on an enemy. These cameras could also come embedded into stickers and other electronics. Phones and cameras could be nano-bugged to transmit all information from the device.

Nanobots could provide basic brainwashing capabilities such as: Destroying your memory and making you feel good or bad about a certain thought. Call it "Positive Reinforcement" if you will. The attacker can deploy this nanobot onto his enemy. He would be able to reprogram his victim while talking about various subjects with him. In a battle situation nanobots could be used against the enemy to make them feel lethargic and loving. A lethargic and loving enemy is not really an enemy.[40]

There is more, so much more to this game-changing new technology. "Guns of all sizes would be far more powerful, and their bullets could be self-guided. Aerospace hardware would be far lighter and higher performance; built with minimal or no metal, it would be much harder to spot on radar. Embedded computers would allow remote activation of any weapon, and more compact power handling would allow greatly improved robotics. These ideas barely scratch the surface of what's possible.

> An important question is whether nanotech weapons would be stabilizing or destabilizing. Nuclear weapons, for example, perhaps can be credited with preventing major wars since their invention. However, nanotech weapons are not very similar to nuclear weapons.[41]

Admiral David E. Jeremiah, Vice-Chairman (ret.), US Joint Chiefs of Staff, in an address at the 1995 Foresight Conference on Molecular Nanotechnology said:

> Military applications of molecular manufacturing have even greater potential than nuclear weapons to radically change the balance of power.

Indeed, nanoweapons would make nuclear weapons as useless as guns made the bow and arrow. Ultimately, the terrifying power of nano-based weapons could make total annihilation of a state's population more possible.

And as *Strategic Trends 2007-2036 Report* makes abjectly clear,

> …it may enable many nations to be globally destructive, eliminate the ability of powerful nations to police the international arena. By making small groups self-sufficient, it can encourage the breakup of existing nations.

The geopolitical predictions of the *Strategic Trends Reports* are no less terrifying: "The exponential growth of molecular nanotechnology (MNT) would destroy the economy, job markets, and much more. The black market would flourish, and countries would break away from alliances under the new power of MNT.

The UN would also cease to have any control over international problems. All of these things would be happening while inevitably waiting for nano-machines to create a doomsday scenario. This scenario is known as an ecophagy (the consumption of an en ecosystem). Imagine a world where self-replicating nano robots consume all matter on earth while exponentially duplicating."[42]

Are the elite playing with fire? Is this part of a long-range plan of demand destruction in order to destroy the world's economy on purpose? If it is, they better be careful of what they wish for.

Do you remember how the "snooping" all started? High tech went from rummaging through your garbage to telephone bugging, and then expanded to Total Information Awareness and Data Mining before going really high tech into PROMIS, nanotechnology and a whole slew of space-age extensions.

TOTAL INFORMATION AWARENESS

Shortly after September 11, 2001, the US Department of Defense proposed a program it called Total Information Awareness (TIA), which would have involved the creation of a huge database containing a substantial collection of private information about every American citizen – all with the objective of catching "terrorists" before they commit the crime.

As in Spielberg's *Minority Report*, fiction mirrors reality.

The data would come from all the electronic tracks we leave behind: financial, housing, education, travel, medical, veterinary, transportation, and communication. At the time, there was an immediate resounding protest across the United States against the program and its violation of civil liberties.

Using the Orwellian tactics of double-speak, the Pentagon briefly changed the name of the program to Terrorism Information Awareness. But then under pressure from American libertarian groups, US Congress took action to prohibit funding for TIA, and the man who championed the project – a convicted felon involved in the Iran-Contra, Reagan-era arms smuggling, retired Admiral John Poindexter resigned.

The new updated, scrubbed, refreshed and made over version of Pentagon's discredited Total Information Awareness is your good, old Data Mining to the *n*th degree: the gradual dissolution of our individual freedoms in favor of a mammoth spy system that is representative of a larger trend that has been underway in the United States and in Europe – the seemingly inexorable drift toward a surveillance society.

In other word, with today's technology, every person and every one of their actions become visible on the world grid, 24/7. Yet, with this erosion of people's Constitutional rights, not a whimper is heard from the beaten down, demoralized, subjugated populace.

UK's Department of Defense defines it as having:

> … complete control of everyone's actions at all time.

This is but a tip of the proverbial iceberg.

PROMIS

What would you do if you possessed software that could think, understand every language in the world, provided peep holes into the innermost secret chambers of everyone else's computers, that could insert data into computers without people's knowledge, enter via the back door of secret bank accounts and then remove the money without leaving a trace; that could fill in blanks beyond human reasoning and also predict what people would do – long before they did it, within a one percent margin of error? You would probably use it, wouldn't you?

Welcome to the world of PROMIS – Prosecutor's Management Information System.

PROMIS is many things to many people. Think of a painting. To the scientist, a snowflake is a snowflake; to an artist, it may be an intricate pattern or an assemblage of minute curved surfaces. On the surface, PROMIS is a product. But it is also something more important and more personal, namely the artist's attitude to the invisible world in general; a question of an attitude of the mind.

It was originally designed to track cases through the legislative, judicial and executive branches by integrating computers of dozens of US Attorneys' offices around the country. The more data you entered, the more accurately the system could analyze and predict the final outcome of the cases.

All information on someone is fed into the software – educational, military, criminal, professional background, credit history, basically anything you can get your hands on, then the software is tasked with making an assessment, and finally rendering a conclusion based on the available information. The more information available, the better the software will predict the outcome.

PROMIS can literally predict human behavior based on information from people. The government and the spooks immediately recognized the financial and military application of PROMIS, especially the National Security Agency, which had millions of bits of intelligence coming into its facilities every day, and with only an antiquated Cray Supercomputer Network to log it, sort it, and analyze it.

In other words, whoever owned PROMIS, once it was fused with artificial intelligence, could accurately predict commodities futures, real estate, or even the movement of entire armies on a battlefield, not to mention every country's purchasing habits, drug habits, stereotypes, and psychological tendencies, in real time –based upon the information fed into it.

The program crossed a threshold in the evolution of computer programming. A quantum leap, if you like. Think of block-modeling social research theory. It describes the same unique vantage point from hypothetical and real life perspectives. For example, pick an actual physical point in space. Now, in your mind, move it further out than you ever thought possible. PROMIS progeny have made possible the positioning of satellites so far out in space that they are untouchable. The ultimate big picture.

There is another advantage: geomatics. The term applies to a related group of sciences – all involving satellite imagery – used to develop geographic information systems, global positioning systems and remote sensing from space. Imaging that can actually determine the locations of natural resources such as oil, precious metals and other commodities.

By providing client nations PROMIS-based software it could then be possible to compile a global database of every marketable natural resource. And it would not be necessary to even touch the resources because commodities and futures markets exist for all of them. An A.I.-enhanced, PROMIS- based program could then be the perfect set up to make billions of dollars in profits by watching and manipulating the world's political climate.

Subsequent research has clearly demonstrated that a similar remote hypothetical position could eliminate randomness from all human activity. Everything would be visible in terms of measurable and predictable patterns. Again, the ultimate big picture.

The other thing to remember is that where mathematics has proved that every human being on the earth is connected to every other by only six degrees of separation, in undercover operations the number shrinks to around three. In the PROMIS arena it often shrinks to two.

PROMIS is not a virus. It has to be installed as a program on the computer systems that you want to penetrate. That's where an Elbit Flash memory chip comes in.

PROMIS maybe fitted with an Elbit Flash memory chip that activates power to the computer when it is turned off. That's because Elbit chips work on ambient electricity in a computer. When combined with another newly developed chip, the Petrie, which is capable of storing up to six months worth of key strokes, it is now possible to burst transmit all of a computer's activity in the middle of the night to a nearby receiver – say in a passing truck or even in a low-flying Signals Intelligence satellite.

There is something else about PROMIS you should know: the trap door. The trap door allows access to the information stored within any database by anyone who knows the correct access code. Sovereign nation's intelligence and banking records can accessed through the Trojan trap door allowing unrestricted access. Which helps ensure the survival of the US hegemony abroad and at home.

Sold to foreign governments, the PROMIS A.I. software could later be accessed by the United States government without other government's knowledge – from the shadows

A goal would be to penetrate every banking system in the world. The elite could then use PROMIS both to predict and to influence the movement of financial markets worldwide.

However, in order to capture each market you would have to control, own or influence each country's Intelligence, Armed Forces, and bank.

But why would one need to go through all the trouble of monitoring all of a foreign country's intelligence operations and military? There's an easier way to get what you want. Fit PROMIS with a Trojan door version on every computer you sell to Canada, Europe and Asia, both civilian and government, to monitor their military, banking and intelligence operations.

This places all data at constant risk of exposure. What makes the whole think spectacular is that many governments are not aware of what they have gotten themselves into. Even if they were, there is little they can do about it at this stage in the game. These are mission-critical systems requiring years of development, not something you whip up in a jiffy at a hot dog stand. Forcing every nation on Earth into cooperating with whoever has the system could easily be done once PROMIS software is on line because software would "control" national banks, intelligence agencies and military.

By cornering, via unlimited access, banking, intelligence and the military – the mere threat of force is all that is needed. A weapon is only good if someone knows what it's capability is. Prior to using the atomic bomb – it was irrelevant.

Governments have been provided with modified PROMIS software that each one of these nations then modified, or thought they modified, again to eliminate the trap door. But unknown to all of them, the Elbit chips in the systems bypassed the trap doors and permitted the transmission of data when everyone thought the computers were turned off and secure. This is how you can cripple everything that Canada, Europe and Asia do that you don't like.

Please understand again, that the good old days of government officials looking through your garbage for clues to your behavior are long gone.

Today, space has become the final frontier.

For example, NASA's unmanned Global Hawk research aircraft developed by Northrop Grumman Corporation flies up to 20,000 meters and provides the longest continuous observation of a tropical cyclone development ever recorded by an aircraft. The images are *captured by Ames Research Center's HDVis camera*. The hurricane surveillance mission is part of the Genesis and Rapid Intensification Process (GRIP) experiment, a NASA Earth science field experiment to better understand how tropical storms form and develop into major hurricanes.

The Global Hawk's science instruments have the capability to peer through cloud tops and measure the internal structure of a storm. The hurricane missions will offer new insights into the fundamental questions of hurricane genesis and intensification.[43]

But, again, it is mission specific and these missions can change and be adjusted to military missions depending on the needs of NASA and by extension, the United States government. That's not all.

Boeing has launched the next generation Global Positioning System (GPS) IIF-1 satellite, the inaugural spacecraft in a 12-satellite constellation that the company is building for the US Air Force.

> GPS is the US Department of Defense's largest satellite constellation, with 30 spacecraft on orbit. The GPS IIF satellites will provide more precise and powerful signals, a longer design life, and many other benefits to nearly 1 billion civilian and military users worldwide.[44]

Raytheon is developing advanced control segment (OCX), which will dramatically affect GPS command, control and mission capabilities and make it easier for the operations team to run the current GPS block II and all future GPS satellites.

Raytheon has over forty years of experience in command and control systems for satellites. Other partners in the project include The Boeing Company, ITT, Braxton Technologies, Infinity Systems Engineering and the Jet Propulsion Laboratory and DARPA.

The Jet Propulsion Laboratory (JPL) in California manages all of the planetary programs for the National Aeronautics and Space Administration (NASA), including ways to make use of laser technology to communicate with spacecraft that are billions of miles away.

A key partner in all these technological military projects is Defense Advanced Research Projects Agency [DARPA]. So, what is DARPA?

According to the DARPA's own web page, it's mission:

> … is to maintain the technological superiority of the US military and prevent technological surprise from harming our national security by sponsoring revolutionary, high-payoff research bridging the gap between fundamental discoveries and their military use.[45]

However, delving deeper, we find DARPA involved in development of a frightening technology with horrific implications. An August 5, 2003 *Boston Globe* story specifically discussed: "Defense Department funding brain-machine work." In it the writer states,

> *It does not take much imagination to see in this the makings of a "Matrix"-like cyberpunk dystopia: chips that impose false memories, machines that scan for wayward thoughts, cognitively augmented government security forces that impose a ruthless order on a recalcitrant population.*

Back in the 1950s:

> DARPA was the dominant sponsor of computer-related research. Cold War-driven projects like SAGE (Semi Automatic Ground Environment), an automated air-defense network of unmanned jet planes, led to a growing interest in war gaming and command systems studies.[46]

Behavioral psychologists like J.C.R. Licklider hoped that:

> … in not too many years, human brains and computing machines will be coupled together very tightly, and that the resulting partnership will think as no human brain has ever thought and process data in a way not approached by the information-handling machines we know today.

"That hope would take form in such later projects as DARPA's Augmented Cognition (Aug-Cog) to create soldier-computer 'dyads,' and the movement for a 'Post-Human Renaissance,' where "there are no demarcations between bodily existence and computer simulation, between cybernetic mechanism and biological organism."[47] This would become the Holy Grail of the front-end research that has spun off not only future battlefield technologies, but also much of today's video game industry."[48]

A 2007 article titled "Video Games and the Wars of the Future" explains this phenomenon in no uncertain terms:

> A 1997 report entitled 'Modeling and Simulation: Linking Entertainment and Defense,' summarized the proceedings of a National Research Council conference which brought together representatives from the military and entertainment world. Their goal was to map out a working relationship whereby the same cutting-edge simulations and virtual reality research brought to bear on enhanced training programs for the military, could also be used in commercially developed video games. Such would be the mission of the Institute for Creative Technologies (ICT)....
>
> The ultimate aim, explicitly outlined by some of ICT's creators, is to actually construct Star Trek's 'holodeck' (the holographic simulations room used on the TV show), whose research includes the role of video-game play on performance in simulated environments: 'Recent neurobiological studies have found that emotional experiences stimulate mechanisms that enhance the creation of long-term memories. Thus, more effective training scenarios can be designed by incorporating key emotional cues.' Creating memories is exactly what simulation research is all about, according to West Point graduate Michael Macedonia, the chief scientist and technical director of PEO STRI who helped create the ICT....
>
> The training techniques being designed by today's visionaries in virtual technologies and artificial intelligence are, in reality, based on nothing more than the reductionist belief that the human mind is a programmable system, not fundamentally different from an animal or machine.[49]

The age of cyborgs, according to DARPA and ICT is just around the corner.

Combined with DARPA's Advanced Wide FOV (field of view) Architectures for Image Reconstruction and Exploitation (AWARE), the system offers the ability to see farther, with higher clarity, and through darkness and/or obscurants is vital to nearly all-military operations.

> The main driver for these requirements is to provide dismounted soldiers, ground troops and near-ground support platforms with the best available imaging tools for their combat effectiveness.
>
> The AWARE program will enable wide FOV, higher resolution and multi-band imaging capability for increased target discrimination and search in all weather day/night conditions, increase operational capability (ability to see panoramic visible scene with multiple target tracking), and provide spectrometry capability using broad band sensors. The AWARE program will solve the current fundamental scaling limitations in imaging systems and demonstrate a design methodology for building compact systems, capable of forming images at or near the full diffraction-limited instantaneous field of view (iFOV) achieved over a wide FOV.[50]

This approach represents a dramatic advance over the current state of technology and allows the government virtual control over every moving object on the face of the Earth. The United States government calls this "Full Spectrum Dominance." Please take a minute and think about the implications.

What's more, DARPA has recently successfully tested gigapixel glass camera with 1.4 and 0.96 gigapixel resolution enabling extremely high resolution shots with smaller system volume and less distortion.[51]

These secret programs, from developing techniques to overcome fundamental limits in current camera scaling, field of view (FOV), pixels and wavelengths continues to advance military imaging across the infrared spectrum. According to DARPA's press release:

Solutions to these limitations will enable high resolution, large FOV, multi-band, and broadband multifunctional camera technologies to enable detection, recognition and identification of targets at longer standoff distances and improve situational awareness.[52]

In November 2012, DARPA unveiled its cutting edge, state of the art, Space Surveillance Telescope, allegedly to: "track and catalogues space debris."[53] Able to search an area in space the size of the United States in seconds, SST:

> ... is capable of detecting a small laser pointer on top of New York City's Empire State Building from a distance equal to Miami, Florida.[54]

SST data will be fed into the Space Surveillance Network, a US Air Force program charged with cataloging and observing space objects. The SSN is a worldwide network of 29 space surveillance sensors, including radar and optical telescopes, both military and civilian.

This information will be used with revolutionary new technology with far reaching military and civilian implications called augmented reality.

Augmented Reality

Augmented Reality is the melting of the real world with the computer-generated imagery. In other words, elements, augmented by computer-generated sensory input, that enrich the user's perception of the real world.

Now, the US Defense Department has ordered special augmented reality contact lenses that create a virtual display superimposed over the normal field of vision. The Defense Advanced Research Projects Agency (DARPA) placed an order with Innovega for lenses, which work with special glasses to enable the wearer to focus on faraway and close objects at the same time. "The human eye on its own can only focus on one distance at a time."[55]

The BBC reports:

> The contact lenses work by allowing the wearer to focus on two things at once – both the information projected

onto the glasses' lenses and the more distant view that can be seen through them – resulting in superhuman vision. They do this by having two different filters. The central part of each lens sends light from the HUD towards the middle of the pupil, while the outer part sends light from the surrounding environment to the pupil's rim.[56]

The system works in conjunction with glasses that project an image or information onto the lens or a display screen. This project has far reaching military applications for DARPA such as enabling soldiers on the ground to see images generated by drones or satellites to complex situational awareness systems encompassing features such as friend/foe identification, multiple sensory interfaces, location intelligence and interactive battlefield medical support, akin to what Arnold Schwarzenegger's character wore in the movie *Terminator*.

Can these technologies be converged on the battlefield? How about a communication network that connects every one of the soldiers and put together an ad-hoc, self-healing mesh network? Every soldiers has a one-two Megabit per second data rate wirelessly, and they will have a six megabit per second rate burst, and because it is ad-hoc self healing, it means all those packets of visual information from video, voice or moving maps are going to jump from soldier to soldier spread over an 18 kilometer area, tailor-made for urban warfare in Mega Cities.

Every time a soldier engaged in combat, their data packets would hop from one to another. The soldier vehicles would have bidirectional amplifiers and antennas in them that could communicate to a command post or an operational center. In other words, for the first time, soldiers in the tactical operations center command post, could see and direct warfare in real time, remotely – *almost as a video gamer.*

Futuristic Technology

Multi-technological convergence has seen us through to the "smart phone" and we are now embracing computer vision. Three-D Augmented Reality has become a gateway to a virtual universe. It allows us to see elements around us, adding an unlimited number of details to what we already know about the objects: the rich, new environment of virtual world.

This new technology can observe a scene and identify the elements within it: people, objects, their history, both past and present. On a simple scale, there are massive opportunities for everyday consumer use. For example, you could browse for hotels and look at the rooms before making a booking.

What ANW (articulated naturality environment) is doing, is to break the barrier that we have with conventional desktop. Current technology only provides us with the 20-30% of the actual digital experience, but with ANW we can get 100% of the experience by embracing that virtual world. ANW opens the door to the virtual universe where our mind is the only boundary.

So, what can be done with all this technology?

DARPA's Video and Image Retrieval and Analysis Tool (VIRAT) and Persistent Stare Exploitation and Analysis System (PerSEAS) programs very soon hope to enable better real-time combat analysis of huge amounts of data generated from multiple types of sensors.

With VIRAT's ability to identify and highlight key actions, and PerSEAS's ability to 'see' dangerous combinations of actions as activities, analysts will soon be able to concentrate on more detailed reviews and understanding of the data.

VIRAT is focused on full-motion video, from platforms such as Predator or Aerostats, allowing analysts to either monitor a live downlink for specific actions of interest, or search an existing archive for past occurrences. These searches may be conducted using a video clip as the input query.

VIRAT finds actions that are short in duration and occur in small geographic areas. PerSEAS focuses on wide-area coverage, such as data from Constant Hawk, Gorgon Stare, ARGUS-IS and other persistent sensors.

> PerSEAS observes multiple actions over a long duration and large geographic regions to postulate complex threat activities. Algorithms from VIRAT provide some of the underlying capabilities within PerSEAS.[57]

There are a number of trends leading to the pervasiveness of these technologies, including: an expanding global economy, the potentially of far-reaching improvements in processing power, a greater cultural assimilation and awareness of technology, and the continued convergence of information and communication technologies (ICT). In turn, according to the *Strategic Trends Report*:

> ICT will itself be a major engine of growth for the global economy.[58]

So, big business of the future is Big Brother's all pervasive control. In fact, with greater and greater population unrest, information communication technologies will be *the big business*. The report agrees:

> It is likely that the majority of the global population will find it difficult to 'turn the outside world off.'[59]

However, a dissociative effect is but one of the aspects of overcoming resistance and changing the established paradigm of society. Listen to Theodore Adorno of Frankfurt School:

> It seems obvious, that the modification of the potentially fascist structure cannot be achieved by psychological means alone. The task is comparable to that of eliminating neurosis, or delinquency, or nationalism from the world. These are products of the total organization of society and are to be changed only as that society is changed.[60]

And what would happen if someone decided to "turn the outside world off" and get off the grid? The world will come looking for you. The *Strategic Trends Report* doesn't mince words:

> Techniques for overcoming resistance can be improved and adapted for use with groups and even for use on a mass scale.

Remember, the two key objectives of the elite are control and population reduction.

Human Cattle

The methodology of taming the captive herd hasn't changed in the past 500 years. In the days of the British Empire, first came gunboats, muskets, and then Venetian-style diplomacy. Thus, the people are subjected, more or less, in the fashion that one herds wild animals into corral.

Then the business of taming the captive herd. Forceful restraint is still obligatory. Those captives tending to rebelliousness must be detected, and either eliminated or reduced to a moral condition of stale jello. The flock must be bred, to evoke in the cultivated descendants the desired attributes of milkiness, meatiness, and docility.

In this way, the captive breed is brought into a state of self-government, in which the ruling bureaucracy is more savagely elitist than the elite themselves. At that latter point in the dumbing-down process, come the winds-of-change, and the captives are entrusted with the duties of fettering themselves at night, or whatever the IMF, the World Bank or the financial markets suggest.

With technological progress and general advancement in all areas of science, the world's population has surpassed seven billion people. With the Internet, we have access to knowledge that half-a-century ago was only accessible to only the elite's highest circles.

A restless ever-expanding-population mass, diminishing natural resources, instant access to information, with the controlled media playing an ever-diminishing role, are all of a great concern to the elite.

We never asked for this. But here we are.

The intense cross-pollination between the computer, biotech and quantum revolutions will bring an unprecedented power to fore in the 21-century. What will the world look like a generation from now?

What's undeniable, the rise of the machine is here. It is all around us. We can no longer ignore it. The augmented man-machine will be faster than us, humans. They will be stronger than us. Certainly, they will last much longer than us.

You may think they are the future. But you are wrong. We are. If I had a wish, I would wish to be human, to know how it feels, to feel, to hope, to despair, to wonder and to love. To know that I am unique through my divine spark of reason. Near future man-machines can achieve immortality by not wearing out. We can achieve immortality simply by doing one great thing ... for humanity.

Endnotes

1 Rudolph Biérent, "Exploring Space: The Optimism Of an Infinite Universe," *EIR*, 23 March, 2012.
2 Helga Zepp LaRouche, "The Next Jump in Evolution: The End of Monetarism," PAC-TV .
3 Ibid.
4 "Mining Helium-3 on the Moon for unlimited energy," *EIR* Volume 14, Number 30, July 31, 1987).
5 "The End of The Obama Administration," LaRouche webcast, *EIR*, Feb 5, 2010.
6 Phil Plait, "Will we ever...live on the Moon," June 14, 2012, BBC).
7 Rudolph Biérent, "Exploring Space: The Optimism Of an Infinite Universe," *EIR*, 23 March, 2012.
8 "The End of The Obama Administration," LaRouche webcast, *EIR*, Feb 5, 2010.
9 Rudolph Biérent, "Exploring Space: The Optimism Of an Infinite Universe," *EIR*, 23 March, 2012.
10 "Mining Helium-3 on the Moon for unlimited energy," *EIR* Volume 14, Number 30, July 31, 1987).
11 "The End of The Obama Administration," LaRouche webcast, *EIR*, Feb 5, 2010.
12 http://goldenspikecompany.com/.
13 http://www.planetaryresources.com/mission/.
14 http://www.bbc.com/future/story/20120613-will-we-ever-live-on-the-moon.
15 http://www.planetaryresources.com/mission/.
16 http://www.wired.com/wiredscience/2012/11/telerobotic-exploration/all/.
17 Dr. Kathie L. Olsen, Ph.D. "FY 2002 Biological and Physical Research Enterprise White Paper," NASA HQ, April 9, 2001.
18 "Almost Being There: Why the Future of Space Exploration Is Not What You Think" http://www.wired.com/wiredscience/2012/11/telerobotic-exploration/all/.
19 http://www.avacore.com/story/be-more-than-you-can-be.
20 http://www.azorobotics.com/News.aspx?newsID=3224.
21 Noah Shachtman, "Be more than you can be," *Wired* magazine, March 2007.

22   Ibid.
23   Terrence Aym, "DARPA creating super bionic insects," Helium.com, May 12, 2012.
24   Ibid.
25   Jonathan Richards, "Military Working on Cyborg Spy Moth," Fox News, May 30, 2007.
26   Ibid.
27   Ibid.
28   Michio Kaku, "Visions of the Future 3/3, The Quantum Revolution," BBC 4.
29   Georgia Miller, "Nanotechnology – a new threat to food," globalresearch.ca, October 30, 2008.
30   http://www.alienscientist.com/metamat.html.
31   Ibid.
32   "Invisibility Breakthrough for Japanese researchers," NTDTV, uploaded to YouTube July 25, 2009.
33   Invisibility (US patent 7256751 Metafractal Cloaking Bands).
34   http://www.alienscientist.com/metamat.html.
35   Michio Kaku, "Visions of the Future 3/3, The Quantum Revolution," BBC 4.
36   Nano Paint Could Make Airplanes Invisible to Radar," *MIT Technology Review,* Katherine Bourzac, December 5, 2011.
37   *DCDC Global Strategic Trends Program 2007-2036*, p.77
38   "Dangers of Molecular Manufacturing," *CRN*, untitled, undated.
39   Kevin Robert Baker, "Nanotechnology weapons," August 3, 2012, http://appreviews4u.com.
40   Ibid.
41   "Dangers of Molecular Manufacturing," *CRN*, untitled, undated
42   Kevin Robert Baker, "Nanotech Weaponry," August 3, 2012.
43   "NASA's unmanned Global Hawk aircraft soars through hurricane surveillance missions," *World Military Forum*, Sept 22, 2010, http://www.armybase.us/2010/09/nasas-unmanned-global-hawk-aircraft-soars-through-hurricane-surveillance-missions/.
44   "The next generation Global Positioning System (GPS) IIF-1 satellite sends 1st signals from space," *World Military Forum*, May 28, 2010.
45   http://www.darpa.mil/mission.html.
46   Oyang Teng, "Video Games and the Wars of the Future," *EIR*, August 10 2007.
47   Tim Lenoir, "All But War Is Simulation: The Military-Entertainment Complex," *Configurations*, Vol. 8, No. 3, Fall 2000, pp. 289-335.
48   Oyang Teng, "Video Games and the Wars of the Future," *EIR*, Aug 10, 2007.
49   Ibid.

50   "Advanced Wide FOV Architectures for Image Reconstruction and Exploitation, http://www.darpa.mil/Our_Work/MTO/Programs/Advanced_Wide_FOV_"Architectures_for_Image_Reconstruction_and_Exploitation_%28AWARE%29.aspx.

51   "DARPA Successfully Tests Gigapixel-class Camera," DARPA, July 5, 2012.

52   "Advanced Infrared Capabilities Enable Today's Warfighter," DARPA, February 21, 2012, http://www.darpa.mil/NewsEvents/Releases/2012/02/21.aspx.

53   "DARPA's Advanced Space Surveillance Telescope Could Be Looking Up From Down Under," DARPA, November 14, 2012.

54   Ibid.

55   Kate Freeman, "These Contact Lenses Give You Superhuman Vision," April 13, 2012.

56   LJ Rich, Dual-focus contact lens prototypes ordered by Pentagon, April 12, 2012, BBC http://www.bbc.co.uk/news/technology-17692256.

57   http://www.darpa.mil/NewsEvents/Releases/2011/2011/06/23_DARPA_advances_video_analysis_tools.aspx.

58   DCDC *Strategic Trends Report*, p.58.

59   Ibíd.

60   Theodor W. Adorno et al., *The Authoritarian Personality*, New York: Harper, 1 950.

Chapter Five

Transhumanism

December 31, 1999: The night before the new millennium. The time of great expectations, of great changes and great unknowns. On that night, the then president Bill Clinton, spoke to America from the nation's capital.

> Tonight we celebrate the change of centuries, the dawning of the new millennium, we celebrate the future. Imagining an even more remarkable XXI century. So we, Americans must not fear change. Instead, let us welcome it, embrace it and create it. Such a triumph will require great efforts from us all. It will require us to stand together against the forces of hatred, bigotry, terror and destruction. It will require us to make further breakthroughs in science and technology. To cure dreaded diseases, to heal broken bodies, unlock secrets from global warming to black holes in the universe.[1]

Then, one day it all changed. 9-11 became the watershed moment in America's and the world's history. Diplomacy ended on September 11, 2001. The reality and risks of an open global conflict, or a more clandestine secret war of attrition were thrust into the immediate and unavoidable focus of a world, which has, for the most part, chosen not to understand what is at stake.

This war will not be fought with bullets and bombs. It is an economic war, a human resources war, a war between nations and their surrogates, a war between secret societies and humanity. Us against them. Seven billion people against a handful of the world's most powerful and ruthless individuals.

The chain of events set in motion on September 11 dictates that the United States, Russia, France, China, Germany, Great Britain, India, Israel and anyone else who can hold his own shall continue on a with series of global military confrontations to control the natural resources of the planet.

With the shedding of the first blood, the dropping of the first bomb, the killing of the first child, the death of the first serviceman; a one-way border to Hell has been crossed. Because of an ever expanding population base, food and water wars are no longer a far-fetched conspiracy, but a frightening reality.

In the process, our civilization has truly reached a point of no return. And with that crossing, economic and political forces have combined to form a "perfect storm"; the ravages of its outburst, now visible in every nook and corner of the planet.

There is no turning back. I wish I could say it another way, soften the blow, somehow. The "New Order" that I describe in this book with such determination and detail is here. It is not a monolith, however; no single group of rich folks sits together in a dark room debating our planetary future. It is, quite literally, an order in which world power aggregates along geographic/geologic lines, forcing *regions* to become players against each other and running roughshod over the nationalist sentiments of their subject populations.

Most people still think in terms of nation-states – a virtuous concept cynically administered by a pathologically meddlesome, promiscuous government spokesmen like a daily dose of vitamins or anti-depressants. I think in terms of money, innovation and technology that money can buy.

Money is a physical concept, not attached to a human sentiment or national identity. That's why all the dots I paint in this book connect in one straight line. It is about a wholesale transfer of the world's wealth into fewer and fewer hands administered by ruthless and increasingly desperate elite.

The collapse of the world's financial system is here, and with it, a crisis of proportions unseen in the annals of human history.

The battle lines have been drawn. Area of conflict: Planet Earth. In the XX century, geography and money proved to be the ultimate trump cards because geography was governing economic decision-making. Afghanistan, Iraq, resources-rich South America and Africa, oil-rich Middle East, slave labor rich South East Asia.

Humanity, throughout history, has faced genocides, atrocities, poverty and starvation. One thing is for certain though, in the end, good has always triumphed – until now. In this new era, technology, like never before, is destined to play a decisive role.

Behind this realignment, enormous streams of capital are being expended and – more importantly – invested behind the scenes. The people controlling this money are not about to see their control dissipate as the nation-states vanish. Money makes its own rules. It is all about *control* … of everything on the planet.

It no longer matters who runs which Western country. The powerful men behind the curtain (secret societies, the elite, London, Wall Street and other financial interests) will remove anyone not to their liking, or bad for their business. Business is money, and money makes its own rules.

Totality of rule is not the only parameter of totalitarianism. The limitlessness of power also proceeds from an omnipresent center. In the new totalitarian movement, this omnipresent directive force communicates through *technological advancement and paradigm shift* – the dominant nodes of the interlocked system.

The Age of Transitions

With the onset of post 9-11 world, we, the people, were asked to make some serious changes, in the name of freedom and fight against terrorists. The war on terror had begun, and with it we found ourselves living in the "terror" of a post 9-11 world.

While people were slapping red, white and blue bumper stickers on their cars, the US government was busy holding a conference concerned with changes so large that they promised to alter human nature itself. XXI century goals were discussed in preparation in what would come to be known as 'The Age of Transitions."

December 3, 2001: A secret conference is held in Arlington, Virginia, USA. The title "Age of Transitions" was coined by Newt Gingrich, in his introduction to the National Science Foundation and Department of Commerce sponsored workshop on NBIC technologies (nanotechnologies, biology, information technology and cognitive technology).

This workshop featured a wide range of participants, from governmental and private institutions to industry and academia such as nuclear and aerospace technology, psychology, computer science, chemistry, venture capital, medicine, bioengineering and social sciences.

It was a chance for experts from NASA, MIT, Carnegie Mellon University, the Department of Defense, Hewlett Packard, the American Enterprise Institute, IBM, Raytheon, DARPA, National Institute of Mental Health and numerous others handpicked for the occasion, to discuss their visions for the future. And the visions discussed for the future were nothing short of Promethean. A stated key goal was "enhancing human performance." This in turn would lead to a "more efficient societal structure."

Technological convergence was given as the answer to solve all of the world's now infamous global problems. I quote from the report:

> Convergence of the sciences can initiate a new renaissance, embodying a holistic view of technology based on transformative tools, the mathematics of complex systems, and unified cause-and-effect understanding of the physical world from the nanoscale to the planetary scale.[2]

The participants at the secret conference promised to bring about a new renaissance of human development. As the world economy descends in a downward spiral to Hell, the time for the elite to make the right decision gets shorter and shorter. They aren't afraid of freedom. They are afraid of the chaos that erupts when individuals have nothing but morality to constrain them. Absolute freedom is no better than chaos.

According to the report:

> Science and technology will increasingly dominate the world, as population, resource exploitation, and potential social conflict grow.[3]

And grow it will.

Technological advancements are not the end of the world, but merely seeds for change.

And change never comes without pain. It's in our nature to want to rise above our limits. Think about it. We were cold, so we harnessed fire. We were weak, so we invented tools. Every time we met an obstacle, mankind used creativity and ingenuity to overcome it. The cycle is inevitable. But, will the outcome always be good? How often have we chased the dream of progress only to see that dream perverted?[4]

The elite, the men who run the world from behind the curtain, have always understood this. Technology offers strength, strength enables dominance, and dominance paves the way for abuse. It also risks giving some men absolute power over others – regardless of the cost to human life.

In the past, we have had to compensate for weaknesses. But what if we get to a point when we never need to feel week or morally conflicted again? What if the paths they, the men behind the curtain, want us to take, enables them to control every aspect of our lives?

The Titan, Prometheus gave mankind fire; a gift he stole from the Gods. Prometheus represents the ability of humankind to discover new forms of technology, new universal physical principles, and to incorporate them into the life and lives of the human species in order to change the relationship of human-beings with the universe as a whole.

Fire, was the first real piece of technology. "100,000 BC – stone tools, 4,000BC – the wheel, IX century AD – gunpowder. XV century – the printing press. XIX century – the light bulb! XX century – the automobile, television, nuclear weapons, spacecraft, Internet. XXI century – biotech, nanotech, fusion and fission, space travel. In the next 50 years, we will be able to create cybernetic individuals, who will be completely indistinguishable from us."[5]

Think about it! Think how far we have come from the first television sets 90 years ago to today, and to the next generation into the future! It is simply mindboggling.

2045: A New Era for Humanity

With that in mind, a new mega-movement was founded – *Russia 2045*. It's objective is to create a new vector for civilization. It is aimed at human development and evolution.

The Global Future 2045 International Congress was held in Moscow in February 2012:

> … laid out a stark vision of the future for neo-humanity where AI, cybernetics, nanotech and other emerging technologies replace mankind – an openly transhumanist vision now being steered by the elite, but which emerged out of the Darwinian-circles directed by the likes of T.H. Huxley and his grandchildren Julian, who coined the term Transhumanism, and Aldous Huxley, author of *Brave New World*. Resistance to this rapid shift in society, the 2045 conference argues, is nothing short of a return to the middle ages.[6]

The elite have great plans for the future: Promethean plans. The downside is that we, the people, have not been invited to the party.

According to Global Future 2045 International Congress:

> … by 2015, we will have an autonomous system providing life support for the brain and allowing it interaction with the environment is created.[7]

By 2020, we will have the ability to transplant the brain into an AVATAR being with AVATAR being, man receives new, expanded life.

> By 2025, new generation of AVATAR provides complete transmission of sensations from all five robotic sensory organs to the operator. By 2030, plans are already in the drawing stages to create 'RE-Brain,' the colossal project of brain-reverse-engineering is implemented. World science comes very close to understanding the principles of consciousness. Scientists are convinced that by 2035, the first successful attempt to transfer one's personality to an alternative carrier will take place. The epoch of cybernetic immortality begins.
>
> In one generation, bodies made of nanorobots, can take any shape or rise alongside hologram bodies. By 2045, we will see drastic changes in social structure. The main priority of this development is spiritual self-improvement. A new era dawns. The era of neo-humanity.[8]

Indeed, the transhumanists have big plans for humanity. Their star-studded goal is called "Project AVATAR" – human like robots controlled via brain computer interface, supported and financed by US Department of Defense via DARPA and NASA.

What's fascinating is that David Cameron's futurist film *Avatar* and our real world are almost identical. The elite plans and the film's storyline are too similar to be a coincidence.

> As *Avatar* begins, the year is 2154. Pandora is run by corporate elite at the top of world government (this is not free-market enterprise but total monopoly). The earth has been mined to depletion its natural world destroyed, and the ruling elite won't hesitate to do the same to other worlds. To facilitate their planned exploitation of Pandora, a scientific elite works under the occupying military force, which in turn serves the mega-corporation financing the mission.
>
> The towering ten-foot blue Avatars are the result of individual human DNA fused to Pandora's humanoid DNA, the Na'vi. Once the hybrid body has been grown in a tank, the team can transfer an individual's consciousness into the avatars, retaining the person's full identity.[9]

This may seem futuristic, but according to a mega project Russia 2045 movement, they intend to:

> Create a new vector for civilization, aimed at future human development and evolution by integrating new discoveries and development from the sciences, physics, energetic, aeronautics, bioengineering, nanotechnology, neurology, cybernetics and cognitive science.

The project is supposed to lead humanity away:

> From the murder of nature and physical death, forward to the realm of freedom and infinite universe of our inner world… war and violence are unacceptable. Man's priority is his development as spiritual self-improvement.

Where have we heard this before? How do you solve these universal problems? By exercising universal control over all of humanity. One World Government anyone?

Russia 2045 mega projects are linked to the *Age of Transitions*, the National Science Foundation Converging Technologies NBIC report, and *Future Trends 2007-2036 Report* issued by the UK government, but discussed several years before at the Bilderberg 2005 conference in Germany.

Henry Kissinger, David Rockefeller, Brzezinski, Hudson Institute, Carnegie Endowment, MIT, Harvard, Stanford, Yale, Bilderberg, CFR, Trilateral Commission: Cross pollination of the same people and organizations pulling the strings from behind the scenes. So, *what will future Man look like?*

By 2015, the race for immortality starts. In the next three years, the transhumanists want to create an AVATAR – a robotic human copy controlled by brain computer interface.

According to *Russia 2045*:

> The 2045 social network for open innovation is expanding with projects such as AVATAR-A, a robotic copy of the human body controlled via Biological Computer Interface. Or an AVATAR-B, in which a human brain is transplanted at the end of one's life. Or an AVATAR-C, in which an artificial brain in which a human personality is transferred at the end of one's life.
>
> Basically, what these scientists want to do is to incrementally move the human mind into more disembodied and, no better way to say it, futuristic vehicles: first a humanoid robot controlled entirely by a human brain via brain-machine interface, then a conscious human brain transplanted into a humanoid robot, then consciousness uploaded to a computer, and finally a hologram that contains a full conscious human mind.[10]

Can this be done? Yet, a bigger question is just how far away is humanity from technically reaching these lofty goals?

Experts agree that a key element is to be able to reverse-engineer the human brain. And the key to reverse-engineering the human brain lies in decoding and simulating the cerebral cortex — the seat of cognition. Futurist Ray Kurzweil and one of the visionaries behind *Russia 2045* project says:

It would be the first step toward creating machines that are more powerful than the human brain. These supercomputers could be networked into a cloud computing architecture to amplify their processing capabilities. Meanwhile, algorithms that power them could get more intelligent. Together these could create the ultimate machine that can help us handle the challenges of the future.

The *Russia 2045* report goes on to mention that: "By 2035, an implantable information chip could be developed and wired directly to the user's brain … synthetic sensory perception beamed directly to the user's senses," and integrated with the global civilization; that is a hive mind, which is outright mind control, with no strings attached, in case you are wondering what I am talking about. But, more on that later.

As we have seen from the *Strategic Trends Report*, disconnecting from the hive mind will get the military police at your door in no time.

Fittingly, two of the attendees at the Bilderberg 2012 conference in Chantilly, Virginia were Anatoly Chubais, CEO of open joint-stock company RUSNANO (formerly Russian Corporation of Nanotechnologies) and a key ally of the Clinton Administration. Chubais is also a former Vice Premier of the Russian government under Yeltsin.

Chubais, in the mid 1990s, was credited with "shock therapy" privatization and the creation of the Russian oligarchs that overnight left 40% of Russians penniless and starving.

The other Russian attending Bilderberg 2012 was Igor Ivanov, Associate member of Russian Academy of Science and President of Russian International Affairs Council, a subsidiary of the powerful US based Council on Foreign Relations.

This is textbook TransHumanism, rooted in many ancient orders and the philosophy of eugenics.

> At its heart, TransHumanism represents an esoteric quest for godhood among certain circles of the elite connected to Masonry, occultism and science/technology wherein supposedly evolving, superior beings 'ethically' replace lesser humans. This philosophy is portrayed in the blockbuster film *Prometheus*, directed by Sir Ridley Scott.[11]

Prometheus the Movie – Future Plans of the Elite

The ideas in *Prometheus* are at the core of many ancient civilizations and the ideas presented in the film at the heart of Western secret societies. Across the world we see early civilizations' obsession with what they believed to be off-world influences.

From Nazca lines in South America to the Pyramids of Egypt, we see artifacts, testament to early men's obsession to off-world manipulators. Every ancient culture believed they were communicating with men from the sky. One could say that *Prometheus* movie is simply art imitating life, and putting a 21-century spin on the beliefs of the Dogon tribe in Africa.

What makes *Prometheus* such a powerful film is the fact that people unquestionably accept the reality presented – always. Popular culture, movies, music carry messages about how society works and how people should behave. Subliminal messages permeate television programs, computer games, magazines, billboards, products, and musical productions. They are just one of the weapons in the arsenal of psychopathic corporations whose entire *modus operandi* concerns profit and dehumanization.

> All that is truly aesthetic, traditional, cultural or substantial merely stands in the way of their monopolistic stratagems, their aggressive, relentless efforts at reducing the entire human race to a body of narcissistic, sense-infatuated, desensitized, amoral, immoral or actively criminal "smiling depressives."[12]

Nevertheless, few people are aware of what we are up against.

> It is a secret, insidious type of war whose main battleground is the people's minds. Its main weapons are propaganda and mass brainwashing mostly by using disinformation, deception, and lies.[13]

This is the real meaning of ideological content. It presents the worldview that influences the people who watch the programming.

Through the use of symbols, Hollywood does it repeatedly, in order to hide real meaning of the story.

Symbols are oracular forms – mysterious patterns creating vortices in the substances of the invisible world. Figures pregnant with an awful power, which, when properly fashioned, in the world of the occult, unleash powerful forces upon the earth. Cynics who continue to doubt whether symbols and images have any lasting negative effect on consciousness should remember and study the most recent example of mass control, that of Nazi Germany. The Nazis openly used ritual choreography, ancient symbols and rallying chants, powerful mystical logos and regalia, and so on, to force men into a hive mentality. The sacredness of numbers begins with the Great First Cause, the One, and ends only with the zero – symbol of the infinite and boundless universe.

Light, the inspired source of illumination, is a key symbol in western secret societies. One such symbol in America is the Statue of Liberty. Statue of Liberty, the original name for which was "Liberty Enlightening the World," signifies occultly the Light Bearer, Lucifer, one of the key symbols of the Illuminati – a fact that French Freemason Frederic Bartholdi, designer of the Statue of Liberty, was well aware of.

The torch is an ancient symbol. In Greek mythology, the original "torch-bearer" was Prometheus, the Titan who stole the divine, impregnating flame from the gods in order to give it to humanity.

In fact, the story of Christianity begins with a great blinding light from which men stumble only gradually, their eyes dazzled, toward the path of spiritual evolution, thus transforming the most perfect of animals on Earth into a potential god.

Which leads to an obvious question: Are we the Gods yet?

Except that the "we" doesn't refer to us, the people, humanity in general, but rather to a privileged elite. 99% of humanity is oblivious to changes taking place around them.

How much of a wild guess would it be, if we were to predict a *dystopian future for mankind where after a financial Armageddon that transferred most of the world's wealth into the hands of the moneyed elite, the poor and wretched masses live under the high tech tyranny of wealthy elite?*

167

The elite have a Promethean plan – to change the world and to transform the very essence of humanity.

The *Russia 2045* conference makes this very clear:

> We are facing a choice to fall into a new dark age, into affliction and degradation, or to find a new model for human development, and create not simply a new civilization but a new mankind. What we need is not a new technological revolution but a new civilizational paradigm. New philosophy and ideology, new ethics, new culture, new psychology and new metaphysics. We must reset our limits. Go beyond ourselves, beyond the Earth and beyond the solar system. New reality and future MAN will arise.[14]

The elite's Promethean goals are two-tiered. On the one hand, the elite have funded technological progress for themselves. On the other, deindustrialization and zero growth as an economic prescription for the rest of mankind.

This was clearly delineated in a report titled "The Changing Images of Man." I quote from the report:

> The concepts of democracy and freedom have disappeared, to be replaced by a high-tech dictatorship based on surveillance, monitoring, mass media indoctrination, police oppression and a radical division of social classes.

Never publicly released and secretly named, "The Changing Images of Man,"[15] sprang from a policy report prepared by the Stanford Research Institute Centre for the study of Social Policy.[16]

The 319 page mimeographed report was prepared by a team of fourteen researchers and directed by a panel of twenty-three controllers including: Margaret Mead, B.F. Skinner, Ervin Laszlo of the United Nations, Sir Geoffrey Vickers of British Intelligence. The entire project was overseen by Professor Willis Harmon, a "futurist," whose specialty was promoting a post-industrial social paradigm as a popular version of how to transform the United States into Aldous Huxley's *Brave New World*.

In a 1961 lecture, Aldous Huxley had described a police state as:

> ... the final revolution" – a "dictatorship without tears," where people "love their servitude.... The twenty-first century will be the era of the World Controllers. He said the goal was to produce "a kind of painless concentration camp for entire societies so that people will, in fact, have their liberties taken away ... but ... will be distracted from any desire to rebel by propaganda or brainwashing ... enhanced by pharmacological methods.... There seems to be no good reason why a thoroughly scientific dictatorship should ever be overthrown.[17]

The aim of the Stanford study, state the authors, is to change the desire of mankind from that of industrial progress to one that embraces "*spiritualism.*" [*War by non-violent or non-linear confrontation*] The study asserts that, in our present society, the "image of industrial and technological man" is obsolete and must be "discarded."

Therefore, the SRI study concludes, we must change the industrial-technological image of man fast:

> Analysis of the nature of contemporary societal problems leads to the conclusion that ... the images of man that dominated the last two centuries will be inadequate for the post-industrial era.

In fact, this far-reaching initiative for silent-weapon technology was first discussed and put into action as an official doctrine by the Policy Committee of the Bilderberg Group at their inaugural meeting in 1954.

Then they used the term "Quiet War" [WWIII] to describe the overt tactical methodology to subjugate the human race. The document, titled *TOP SECRET: Silent Weapons for Quiet Wars, An introductory Programming Manual,* was uncovered quite by accident on July 7, 1986, when an employee of the Boeing Aircraft Company purchased a surplus IBM copier for scrap parts at a sale and discovered inside details of a plan, hatched in the embryonic days of the "Cold War." This manual of strategies called for control of the masses through manipulation of industry, peoples' pastimes, education and political leanings. It called for a quiet revolution, pitting brother against brother, to divert the public's attention from what is really going on.

Here is a partial quote from this document (TM-SW7905.1):

> It is patently impossible to discuss social engineering or the automation of a society, i.e., the engineering of social automation systems (silent weapons) on a national or worldwide scale without implying extensive objectives of social control and destruction of human life, i.e., slavery and genocide. This manual is in itself an analog declaration of intent. Such writing must be secured from public scrutiny. Otherwise, it might be recognized as a technically formal declaration of domestic war. Furthermore, whenever any person or group of persons in a position of great power and without full knowledge and consent of the public, uses such knowledge and methodology for economic conquest – it must be understood that a state of domestic warfare exists between said person or group of persons and the public. The solution of today's problems requires an approach, which is ruthlessly candid, with no agonizing over religious, moral or cultural values. You have qualified for this project because of your ability to look at human society with cold objectivity, and yet analyze and discuss your observations and conclusions with others of similar intellectual capacity without a loss of discretion or humility. Such virtues are exercised in your own best interest. Do not deviate from them.

The Future is We

It was the hope of the three conferences, "The Age of Transitions," the "Global Future 2045 International Congress" and the "NBIC" conference, to integrate humanity with nature to save the Planet Earth from mankind. Visions laid out included robotics, cybernetics, artificial intelligence, life extension, brain enhancement, brain-to-brain interaction, virtual reality, genetic engineering, teleportation, human-machine interfaces, neuromorphic engineering and enhanced human capabilities for defense purposes.

One thing is obvious. For the first time in history, the elite feel they have a chance to steal the fire from the gods. To turn away from it now, to stop pursuing a future in which technology and biology combine, leading to the promise of a "Singularity" – would mean to deny the very essence of who they are.

The Singularity is an era in which the collective intelligence of mankind will become increasingly, "nonbiological and trillions of times more powerful than it is today – the dawning of a new civilization through technological means,"[18] which will enable us to transcend our biological limitations and amplify our creativity. I will come back to the idea of Singularity later in the chapter.

The elite can now become the gods they have always been striving to be. We, the great unwashed, might as well get used to it.

After everything they have seen, all the fighting and the chaos, the fact is, as far as the elite are concerned, deciding the future for all of mankind shouldn't be left up to a mass of raw minds swayed by elementary needs.

The elite's actual affection for the masses is captured in the following passage by T. H. Huxley:

> The great mass of mankind has neither the liking, nor the aptitude, for either literacy, or scientific, or artistic pursuits; nor, indeed, for excellence of any sort.

And in any case, he said, the "great mass" was doomed to poverty due to overpopulation:

> What profits it to the human Prometheus, if the vulture of pauperism is eternally to tear his very vitals?

This is the very same sentiment that another secret society, Britain's Coefficients Club, expressed back in 1902. The Bilderberg Group, in fact, is a natural extrapolation from this Club, founded in 1902. Lord Alfred Milner of the Round Table spoke of his vision for the future during a 1903 meeting at St. Ermin's Hotel, over half a century before the Bilderberg Group was founded.

Milner stressed a point:

> We must have an aristocracy – not of privilege, but of understanding and purpose – or mankind will fail. The solution does not lie in direct confrontation. We can defeat democracy because we understand the workings of the human mind, the mental hinterlands hidden behind

the persona. I see human progress, not as the spontaneous product of crowds of raw minds swayed by elementary needs, but as a natural but elaborate result of intricate human interdependencies.

In the early XX century, the common sentiment of the elite was called eugenics. After World War II with genocide still fresh in people's minds, the term was first changed to "interdependence," and later in the century to "reengineering."

Today, it is called convergence – turning the genocidal view of the world upside-down and inside out. Convergence is a priority area of importance in implementing *the great promise of a new day* for the XXI century.

In the past half a century we have unlocked some of the greatest secrets of cosmos, and have advanced mankind towards a new world, breathing new fire and spirit into our understanding of the Universe. And in the process, some of the ideas have gone from being awe-inspiring and revolutionary in their content to being perceived as part of the conventional wisdom.

Transhumanism

One generously funded and organized group stands above all others in plotting this funding for convergence – the "World Transhumanist Association." Transhumanism is an ultra high tech dream of computer scientists, philosophers, neural scientists and many others. It seeks to use radical advances in technology to augment the human body, mind and ultimately the entire human experience. It is the philosophy that supports the idea that mankind should pro-actively enhance itself and steer the course of its own evolution.

Transhumanists wish to become what they call "post-human." *A post-human is someone who has been modified with performance enhancing body and brain augmentations to the point that they could no longer call themselves human.* They have mutated themselves into an all-together new being.

Many people have trouble understanding what the true transhumanism movement is all about, and why it's so awful.

After all, it's just about improving our quality of life, right? Or is transhumanism about social control on a gigantic scale?

In fact, "social control" is exactly how powerful American foundations such as the Rockefeller Foundation, Carnegie Endowment and Macy Foundation see it. The ability to make machines act like humans, and the ability to treat humans as machines:

> … the final accomplishment of H.G. Wells' old Fabian goal of a 'scientific world order' where everything is as neat as a differential equation, and unpredictable things such as human creativity never mess things up.[19]

To most people, this sounds like something from a science fiction film. Few are aware of constant breakthroughs in technology, which makes the transhumanist vision a very real possibility for the near future.

For example, neurochip interfaces, computer chips that connect directly to the brain are being developed right now. The ultimate goal of a brain chip would be to increase intelligence thousands of time over – basically turning the human brain into the super computer.

Lifelong emotional well-being is also a key concept within transhumanism. This can be achieved through a recalibration of the pleasure centers in the brain. Pharmaceutical mood renders have been suggested, which will be cleaner and safer than the mind-altering drugs.

This is Huxley's XXI century scientific dictatorship without tears. The era of the World Controllers … and as Huxley said,

> There seems to be no good reason why a thoroughly scientific dictatorship should ever be overthrown.

The goal is to replace all aversive experience with pleasure beyond the doubts of normal, human experience. Nanotechnology, for example, is a pivotal area concerning transhumanists. It is the science of creating machines, which are the size of molecules. Such machines could create organic tissue for medical use.

Using this type of technology could dramatically prolong life span. The Global Future 2045 International Congress predicted that by 2045, it will be possible to live forever by combining technology and biology into an event called: "The Singularity."

The Singularity would occur at the point in which artificial intelligence surpasses the capabilities of the human brain. From cyborgs with very long life spans, to downloading consciousness itself into a machine, transhumanists say that it is impossible to predict exactly what a post-human will be, *but that it will be better*. Such lofty promises are being embraced by many people – for a better world for everyone.

No matter how you look at it, Singularity is being promoted as *the* great solution to our XXI century global problems. And thanks to the human genome project, we will soon be able to decode DNA itself. Through the use of applied genetics, science will then be able to improve the human race. What most people don't realize is that this concept is not new.

Transhumanism was born out of humanism, which is yet another clever disguise of "scientism," created specifically so that global eugenics operations could be carried out without being noticed.

Eugenics

Throughout history, there have been those, who, with an eye to specific political objectives have used terror or the threat of terror against targeted populations. The scientific rationale for tyranny has always been attractive to the elites, because it creates a convenient excuse to treat their fellow men lower than animals.

Eugenics, a crackpot notion of hereditary superiority and inferiority, originated in the 1880s and 1890s, and spawned by a British network of families, including Darwin's cousin, Sir Francis Galton, Thomas Huxley, Sir Arthur Balfour, the Cadbury and the Wedgewood families as well as other late XIX century British Empire strategists linked to the Round Table movement of Cecil Rhodes and Lord Alfred Milner.

> Darwin and Huxley were members of the networks set up by the British East Indian Company and Privy Council to remold the cultural, scientific, and religious philosophy in England for imperial rule. Since Charles Darwin virtually

never spoke in public, Huxley became his mouthpiece, his self-proclaimed 'bulldog.'[20]

The name "eugenics" was coined by Galton from a Greek term meaning "wellborn," and already in 1869 Galton had written a book, *Hereditary Genius*, which argued that the aristocratic families of the British Empire were in fact, a superior race. That mental qualities are biologically inherited; that the white race is the biologically best endowed to dominate the world, and that the English are the cream of the white race. The fact that in the struggle for life, they had made it to the very top of society proved that they were the very best that humanity had to offer.

In his 1872 *Enigmas of Life*, W.R. Greg, considered the co-founder of eugenics with Galton, said that Britain:

> ... owes her world-wide dominion and ... the wide diffusion of her race over the globe, to a daring and persistent energy with which no other variety of mankind is so largely dowered.... At all events it is ... the strongest and the fittest who most prevail, multiply, and spread, and become in the largest measure the progenitors of future nations.[21]

Darwin credited his discovery of evolution to Reverend Parson Thomas Malthus, a hired pen for the British East Indian Company, who popularized the theory of: "scares, limited natural resources." Malthusian law is similar to what was proposed for the UN 1994 Cairo Population Conference: a theory in demographics regarding population growth developed during the industrial revolution on the basis of the writings of Malthus' famous 1798, *An Essay on the Principle of Population,* which was nothing more than a plagiarized version of Venetian monk Giammaria Ortes' 1790 publication, *Riflessioni sulla popolazione delle nazioni*. According to his theory, population expands faster than food supplies.

That is the typical sophism that the British Empire, throughout its history, has attempted to use to brainwash human populations into believing in the existence of the limits to growth. But the idea of sustainability capacity is based on animal models.

Hence the idea that for a certain square kilometers of land, for a given surface area, there is only a limited number of individuals of that species that can be sustained or maintained by that territory.

Malthus presented these ideas saying that human populations grow at an exponential rate or a geometric rate. They multiply by two every thirty or forty years, and that the ability of land to sustain these populations producing food, increases arithmetically. Through this calculation, Malthus stated that there was a limit of sustainability for the land around the planet in terms of human populations. And that soon, we as people were going to exceed this limit sustainability.

Darwin emphasized his dependence on Malthus right in the introduction to his 1859 book, *The Origin of Species*, whose full title is *On the Origin of Species by Means of Natural Selection, or the Preservation of Favored Races in the Struggle for Life.*

> The Struggle for Existence amongst all organic beings throughout the world ... inevitably follows from their high geometrical powers of increase…. This is the doctrine of Malthus, applied to the whole animal and vegetable kingdoms. As many more individuals of each species are born than can possibly survive; and as, consequently, there is a frequently recurring struggle for existence, it follows that any being, if it vary however slightly in any manner profitable to itself … will have a better chance of surviving, and thus be *naturally selected.*

Natural selection is genocide, extrapolated to its logical conclusion, a time worn theme for the world's elite. According to its twisted logic, some people deserve to live, but most of us, deserve to die. Darwin himself stated:

> … elite status is prima facie evidence of evolutionary superiority.

Not surprisingly, the Royal Society, the scientific institution dedicated to the improvement of natural knowledge picked up on these new ideas, and promoted Darwin heavily. Being the creation of the British Monarchy, the Royal Society was obviously in favor of promoting the idea of the genetic superiority of the Royal Family.

Science was positioned to replace the old religious appeal of the Divine right of Kings to rule over inferior races – us.

Malthus and the British East India Company

Malthus was not just any old country parson, but the official chief economist for the British East India Company (BEIC), the largest monopoly the world had ever seen, with an army in the late 18th and early 19th centuries that was larger than that of the British government itself. In fact, the slave-trading and dope-pushing BEIC *was* the British Empire. And when the BEIC set up its Haileybury College in 1805 to train its officials, they appointed Malthus as the very first professor of political economy in Britain, actually in the world. Malthus's students over the next several decades became the BEIC's administrators, and systematically applied his policies of genocide to keep the native populations under control. They killed tens of millions in India alone, including by forcing them to grow opium instead of food, which opium the BEIC then used to poison the Chinese.[22]

For example, between 1770 and 1771, five years after the consolidation of the rule of the British East India Company on the Indian sub-continent, ten million Indians died of hunger and malnutrition. The British East India Company was at that time aspiring to World Empire. They had not only consolidated their control of the British monarchy but also of Parliament. They had full control of British politics at the time. The British had just defeated the French in the Seven Years War, and the British East India Company was committed to an imperial policy that was made evident with the first Bengal famine in 1770.

But the genocide in India did not stop there. In 1783, eleven million Indians died in less than two years, in the famous Chalisa famine. Some regions of India were completely depopulated, 60% of the villages and towns disappeared from the face of the earth.

Five years later, in 1788-89, some eleven million Indians died in the "Famine of the Skull," named for the bleached bones of the victims whose bodies were piled up by the tens of thousands in the gutters. The living could not cope with burying such quantity of dead people.

However, India was but one example of the "natural selection process" executed by London.

The famous case is that of the potato famine in Ireland, called the "Great Hunger," another example of organized genocide. It was well known in Ireland that there was not a famine. Famine means food shortages. The Great Hunger was caused by British Imperialist policy.

There was plenty of food but it was being systematically exported during this famine. In 1847, two million Irish people starved and an estimated two million others emigrated. This means that between starvation and emigration, the population of Ireland was cut in half, from 8.5 million people to just over 4 million in two short years. This falls under the definition of *genocide*.

The current situation in Africa should be on everyone's mind because we are talking about planned depopulation. What have British politics achieved after decades of colonial rule? The denial of technology.

How do you explain that most of Africa is a famine invested no-man's land of populations living in sixteenth and seventeenth centuries' conditions, when as civilization, we have access to nuclear energy, desalination and high speed rail? Remember, we are in the year 2013. Massive diseases, technology denial, forced sterilization of the populations – this is what the real Africa looks like.

Forget about the UN based propaganda. This is a direct result of the racist imperial ideology – the idea that some races of the world are less deserving than others to having access to modern technology and modern living standards.

Darwin and the X Club

Darwin, a neurotic hypochondriac who rarely left his house,

> … was not a man, but a project, a figurehead for the cultural warfare that was run top-down by the Privy Council of the British Crown through the British East India Company and its network of salons and front-groups, such as the Metaphysical Society, the Oxford Essayists, the Coefficients Club, the Cambridge Apostles, who coined the term 'agnosticism, and the elite men's clubs of London, including the X Club of so-called scientists, which Huxley founded on November 3, 1864 to ram through Darwinism.

Huxley was a leading figure in the so-called Working Men's Movement, which was actually founded by the elite of Cambridge University, just like its successor of a couple of decades later, the Fabian Society. He lectured to these early socialists on Darwinism and 'modern scientific method.[23]

According to a history of the club:

> … the X Club can be regarded as the cabinet of a liberal party in science. Its policies were to advance research, to reform the public image of science, and to disseminate science and scientific attitudes in society. From 1860-1890 it was influential. It was the party in power between 1870 and 1885. Under the leadership of the X Club science became central to English culture.[24]

The Notion of Evolution

The very notion of 'evolution,' which Darwin supposedly invented, had already been proposed by others. His grandfather Erasmus Darwin, for instance, had proposed 'common descent' in his 1794 book *Zoonomia*. While Darwin's famous Tree of Life diagram, showing 'common descent,' with all species being derived from one or a handful of original primitive species, had already been published in a less elaborate form in a famous 1844 book by Robert Chambers, *Vestiges of the Natural History of Creation*. As for the idea that one species evolves into another species due to small changes in individuals within a species, that idea of 'transmutation of species' was put forward by the French naturalist Jean-Baptiste Lamarck in his 1809 book, *Philosophie Zoologique*. The theory of 'natural selection,' the presumed engine of evolution, had been presented to the Royal Society in 1813 by Dr. William Charles Wells, who fled America for England at the outbreak of the American Revolution.[25]

Eugenics in America

In the United States, the story of eugenics begins in 1904, when the Cold Spring Harbor Laboratory was started by prominent eugenicist Charles Davenport with the funding of leading American oligarchs, the Rockefellers, Carnegies, Harrimans, J.P. Morgan Jr., Mary Duke Biddle of the tobacco family, Cleveland Dodge, John Harvey Kellogg of the breakfast cereal fortune, and Clarence Gamble of Proctor & Gamble.

All of them were quietly funding eugenics as members of the American Eugenics Society, and funding experiments into the forced sterilization of "inferior people" and various forms of population control as far as World War I.

> Millions of index cards on the bloodlines of ordinary Americans were gathered, to plan the possible removal of entire blood lines deemed inferior. The aim of the index card project was to map the inferior bloodlines and subject them to lifelong segregation and sterilization to kill their bloodlines.[26]

By 1910, the British had created the first network of social workers expressly to serve as spies and enforcers of the eugenics race cult, which was rapidly taking control of Western society. Winston Churchill, economist John Maynard Keynes, Arthur Lord Balfour and Julian Huxley, who went on to be the first head of UNESCO after the war, were all devout eugenicists.

Clearly, one of the causes of world problems, as the elite saw it then and now, was:

The persistent tendency of the human species to reproduce and multiply itself. An increasing number of people in the world meant a greater potential to cause trouble and to demand a bigger slice of the Big Pie of Life, which the Rockefellers and their wealthy friends regarded exclusively as their God-given right.[27]

However, Hitler's British financial backers weren't the only ones sponsoring eugenics research. In the 1920s, the Rockefeller family bankrolled the Kaiser Wilhelm Institute for Genealogy and Demography,[28] which later would form a central pillar in the Third Reich.

At the end of the war, with corpses still smoldering across Europe, the allies protected from prosecution the very Nazi scientists such as Joseph Mengela, who had tortured tens of thousands of people to death. The Nazi radical brand of eugenics had embarrassed the Anglo-American social controllers, making "eugenics" and "mental hygiene" dirty words.

The controllers, however, wouldn't be deterred. In 1956, the British Eugenics Society decided in a resolution that, "the society should pursue eugenics by less obvious means." This meant "planned parenthood" and the environmentalist movement. The World Wildlife Fund and its direct action terrorist arm, Greenpeace, as well as like-minded groups are not just a lunatic fringe that can easily be ignored; they are the shock troops of the oligarchy in their fight against humanity.

Every population control policy has been simply renamed as they continued their work under the protection of the United Nations and its associate organizations. The Eugenics, Euthanasia and Mental Hygiene Societies of Britain, America and the rest of Europe, was simply renamed to the more palatable Mental Health Association of Great Britain, and National Association of Mental Health of the United States that later became World Federation of Mental Health.

Eugenics Quarterly Magazine became *Social Biology,* and the American Birth Control League became Planned Parenthood, which today is responsible for a massive depopulation in Africa.

It is not widely known that some of the largest aid agencies, and U.S. Christian Fundamentalist groups, have been covertly running pogroms in Africa over the past several decades.

Its banner is "Family Planning," which is being turned on its head, when one fully understands the real implications and far-reaching objectives. These family planning policies are vigorously and consistently advocated by major bilateral donors such as the U.S. Government through its surrogate, U.S.-AID and multilateral agencies, most notably the International Planned Parenthood Federation (IPPF), the U.N. Fund for Population Activities (UNFPA) and the World Bank in Africa.

Eugenics was being accepted as a genuine form of science. Social Darwinism made strong advances towards a world in which scientism would fulfill Galton's dream of having eugenics be the religion of the future.

Prominent eugenicist Julian Huxley stepped up and offered a solution. He simply invented a new word to replace eugenics. That term being, transhumanism, which he defined as:

> A need for mankind to realize the importance of steering the direction of its own evolution.

Yes, eugenics was one of the original aspects of transhumanism, and it was no surprise that Julian was a President of British Eugenics Society, which had the task of removing undesirable "variants" from the human gene pool.

Meanwhile, immediately after the war, Sir Julian Huxley changed the name of the program for enforced birth control, zero economic growth, the technology of mass mind control, and continued to apply the principles which created Nazi Germany's mass genocide against the "racially unfit."

In 1946, Huxley announced that:

> Even though it is quite true that any radical eugenic policy will be for many years politically and psychologically impossible, it will be important for UNESCO to see that the eugenic problem is examined with the greatest care and that the public mind is informed of the issues at stake so that much that now is unthinkable may at least become thinkable.

In 1974, Henry Kissinger, Huxley's intellectual son said,

> Depopulation should be the highest priority of foreign policy towards the third world.

Huxley was a founder of the British Eugenics Society, and the first Director-General of UNESCO, which pushed for population reduction, and what Huxley called "a single culture for the world." He was also a leading member of World Federation of Mental Health, director of the Abortion Law Reform Association, and founder of the World Wildlife Fund, whose first president was the former card-carrying Nazi, Prince Bernhard of the Netherlands. Prince Bernhard was also one of the organizers of the Bilderberg meetings since its inception in 1954.

Few realize that Huxley represented a social set of miscreants the likes of which have not been seen since. Amongst them was his grandfather, Thomas Huxley, whose outspoken support of Charles Darwin's theory of evolution earned him the nickname, 'Darwin's Bulldog; his personal mentor and Satanist Aleister Crowley; Brigadier John Rawlings Rees; Bertrand Russell and H.G. Wells, whose Open Conspiracy was a call for fascist imperial world dictatorship.

> What makes the "open conspiracy" open, is not the laying out of some secret master-plan, not the revealing of the membership roster of some inner sanctum of the rich and powerful, but rather, the understanding that ideas, philosophy and culture, control history.

Also involved is Julian's brother, Aldous, of the *Brave New World* fame. In Aldous Huxley's book, the population is genetically divided into Alphas, Betas, all the way down to the synthetically produced Epsilon morons through the application of eugenics, or Nazi breeding laws.

Hive Mind

The transhumanists have a popular term – "hive mind," which refers to the giant collective intelligence that might be created when people all over the world link their brains together with technology. In other words, creating a whole new intelligence through symbiotic existence. In reality, the hive mind will be Galton Darwin's beehive, a creation of a new man, which has been written about for many years.

There is a great deal of talk about the hive mind in the NBIC report. I quote:

> We envision the bond of humanity driven by an interconnected virtual brain of the Earth's communities searching for intellectual comprehension and conquest of nature.
>
> A networked society of billions of human beings could be as complex compared to an individual human being as a human being is to a single nerve cell. From local groups of linked enhanced individuals to a global collective intelligence, key new capabilities would arise from relationships arising from NBIC technologies. Such a system would have distributed information and control and new patterns of manufacturing, economic activity, and education. It could be structured to enhance individuals' creativity and independence. Far from unnatural, such a collective social system may be compared to a larger form of a biological organism. Biological organisms themselves make use of many structures such as bones and circulatory system. The networked society enabled through NBIC convergence could explore new pathways in societal structures, in an increasingly complex system.[29]
>
> It may be possible to develop a predictive science of society and to apply advanced corrective actions, based on the convergence ideas of NBIC. Human culture and human physiology may undergo rapid evolution, intertwining like the twin strands of DNA, hopefully guided by analytic science as well as traditional wisdom. The pace of change is accelerating, and scientific convergence may be a watershed in history to rank with the invention of agriculture and the Industrial Revolution.[30]

It is this human beehive, which has been the ideal society in the eyes of the elite for a very long time. This is a template for "post humanity," the ultimate slave race, scientifically designed to never rebel, and the wholesale disappearance of the human being with his and her divine spark of reason at the expense of group speak, group think and group actions.

Julian Huxley could not have said it better himself. However, let's not forget Julian's brother, Aldous Huxley, author of *Doors of Perception* and the *Brave New World*. On March 20 1962, he gave a lecture at the UC Berkley in which he clearly laid out the vision of a planned future society:

> In the past, we can say that all revolutions have essentially aimed at changing the environment in order to change the individual. Today, we are faced with the approach of what may be called the ultimate revolution, the final revolution where man can act directly on the mind body of his fellows. The nature of the ultimate revolution which we are now faced is precisely this; that we are in process of developing a whole series of techniques which will enable the controlling oligarchy that always existed and presumably will always exist, to get people to actually love their servitude.
>
> First of all, to standardize the population, to iron out inconvenient human differences. To create mass produced models of human beings, arranged in some kind of a scientific class system. The number of predictions which were purely fantastic when I made them 30 years ago, have come true or are in the process of coming true; not through terror but through making life much more enjoyable than it normally does. Enjoyable to the point where human beings come to love the state of things that by any reasonable human standard they ought not to love. And this, I think, is perfectly possible. One of the more recent developments in the sphere of neurology is the implantation of electrodes in the brain. This of course is being done on a large scale in the behavior of rats.

Returning to the NBIC report:

> A vast opportunity is created by the convergence of sciences and technologies starting with integration from the nanoscale, having immense individual, societal, and historical implications for human development. Therefore, the

contributors to this report recommend 'a national research and development priority area on converging technologies focused on enhancing human performance.' Advancing knowledge and transforming tools will move our activities from simple repetitions to creative, innovative acts and transfer the focus from machines to human development. Converging technologies are at the confluence of key disciplines and areas of application, and the role of government is important because no other participant can cover the breadth and level of required collective effort. Without special efforts for coordination and integration, the path of science might not lead to the fundamental unification envisioned here. Technology will increasingly dominate the world, as population, resource exploitation, and potential social conflict grow. Therefore, the success of this convergent technologies priority area is essential to the future of humanity.

Transhumanism fills people's hopes and minds with dreams of becoming superhuman, but the fact of the matter is that the true goal is the removal of that pesky, human free will itself.

There is no possibility for discovery of a new universal principle of science because the actual principles of human knowledge are not confined to what is learned. As Johannes Kepler exhaustively explained in his, *The Harmonies of the World*: "the human personality has access, by the unique nature of human mind's unique power of cognition," unique among all living creatures, "to generate previously unknown, validatable universal physical and comparable principles" which increase our species' power in and over the universe per capita and per square feet of space.

To annul this unique nature of human mind, a parallel world of Jungian thought can now be found to be intertwined with the ideas of "New Age." From beer commercial memes that talk of the collective unconscious, to the Star Wars series and the mass-marketed virtual reality, these thoughts are consciously building on Jungian imagery and mystical ideas. To them:

> … this new technology is the key to open the doors of perception to Jungian dream world.[31]

However, this Jungian Star Wars "dream world" is more like a recurring nightmare. The most crucial epistemological issue is often overlooked. Do these Star Wars, Doom and Lord of the Rings creatures look human to you? The second issue is how do you corrupt people into Jungian mysticism? Obviously, by dehumanizing the image of man. Therein lies a principle of evil pervading "virtual consciousness."

Post-humanity will be a new human, which has been genetically engineered and brain-chipped for total control. Part man and part machine, the new man will no longer have a need of the sexual reproductive function. If the elite's plan is to reduce the world's population, can you think of a better way to do that?

What does the NBIC report says? Exactly that.

> Individuals have served to keep the gene pool stirred up and healthy via sexual reproduction, but this data handling process would no longer necessarily be linked to individuals. With knowledge no longer encapsulated in individuals, the distinction between individuals and the entirety of humanity would blur.

This may seem like something out of a science fiction novel, but this bizarre and horrific scenario is not only possible, it is intended, although we won't arrive at this point overnight. But how?

Virtual Reality

First, we must step into the world of virtual reality, where our identity as purely human may blend with that of our new virtual persona, where you can be someone else. Is this possible?

Few are aware of constant breakthroughs in technology, which makes the transhumanist vision a reality in the next five years. Artificial intelligence, and the creation of thinking robots is closely related to the mind-machine-merger concept of neurochips interfaces. Computer chips that connect directly to the brain are being developed right now. The ultimate goal of a brain chip will be to increase intelligence thousands of time over, basically turning the human brain into the super computer.

There is a growing perception that machine parts may be added to human bodies to create cyborgs, a term invented by Dr. Nathan Kline, a psychiatrist linked to the CIA and MK-ULTRA, mind control experiments.

Kline had co-written an article with Manfred Clynes for the September 1960 issue of *Astronautics* entitled: "Cyborgs and Space," which first introduced the term "cyborg" into the English language.

Using the problem of deep space travel, they discussed "the cybernetic aspects" of homeostatic processes in the body in order to develop ways to keep astronauts awake for weeks.[32]

Lifelong emotional well being is a key concept within transhumanism. This can be achieved through a recalibration of the pleasure centers in the brain. The building of a virtual utopia, via mood renders, and the merging of nanobots and biology.

> If you can use nanobots to control your brain, you wouldn't even need money and possessions to be happy. You could stimulate any living experience in your head.[33]

The goal is to replace all aversive experience with pleasure, beyond the doubts of a normal human experience.

Nanotechnology, for example, is a pivotal area for transhumanists. It is the science of creating machines, which are the size of molecules. Such machines could create organic tissue for medical use. Using this type of technology could dramatically prolong life span. For how long, you ask? How does forever sound to you?

In less than one generation, humanity may merge totally with technology by uploading individual consciousness to virtual reality. Upon being uploaded, one could then live forever within a computer-generated reality – leaving the physical body behind.

In this machine, the individual could merge his intelligence with the collective intelligence of all others in the digital reality, effectively becoming one super intelligent being. A concept we have already discussed: the 'hive mind.'

But, it goes way beyond that. One of the grand visions is to upload human minds into holographic bodies. A three-dimensional hologram, blurring reality and virtual reality by projecting images with blue beam or barium and strontium (contents of nuclear fallout), turning our beloved blue sky into a giant mind-numbing television set. You don't believe me?

Back in 1998, U.S. Army Research Laboratory patented a three-dimensional holographic display using strontium and barium for holography. In fact, the US government has a patent for holographic clouds (Patent 5409379), which was filed way back in 1993. In 1994, the USG patented the use of holographic clouds in a scene (Patent US 5489211).

Again, how many people see a parallel with Jung's 'collective unconscious'?

> Only through the recognition of the dimension of the collective unconscious can science serve the interests of man.

Except that virtual reality and computers cannot replicate human intelligent thought process. What AI can do, is to alter consciousness, in a similar way that LSD played with the human mind back in the 1960s.

The World of Matrix

The government and industry has done a good job inserting us into a Matrix. We are being sucked into our iPhones, sucked into video games, into RD interface. Theoretical physicist, Michio Kaku is convinced that: "by 2020 there will be an entire three dimensional universe in cyberspace, with virtual countries and governments, virtual schools and universities, virtual property and stock markets, and virtual family and friends."

According to a futurist, Ray Kurzweil:

> Virtual reality will become more and more like real reality but have the advantage that I can share a virtual reality environment, incorporating all of the senses, with someone else, even if they are hundreds of kilometers apart. It has a lot of advantages over real reality.

There is a general consensus amongst people that:

> … in 10 years, things like second life will become just as prevalent as email now, and I think virtual worlds will have a similar way for people to get together and communicate.

To give skeptics the benefit of the doubt, perhaps this version of humanity is pure speculation.

However, converging technology does present a need for radical change within society, no matter how it is applied. This point is made over and over again in the National Science Foundation NBIC 2001 report.

The message of the report is abjectly clear. *The world of virtual reality is a world of totally controlled environment.*

Jaron Lanier of VPL Research, Inc. says:

> If you can generate enough stimuli outside one's sense organs to indicate the existence of a particular alternate world, then a person's nervous system will kick into gear and treat the stimulated world as real.

Furthermore, if virtual reality technology is administered as a brainwashing tool, it can weaken what the Freudians call the "super-ego," that part of the personality that coheres to one's moral conscience, which is the basis for truth seeking on the one hand and uniquely human act of fundamental discovery and integration of new universal principles which have improved the lives of people per square feet of space against nature on the other.

In fact, science has tried to phase out our built in connection to the universe, and replace it with the "consensual hallucination" of virtual reality.

Please understand this environment that has been created, made-to-order by the corporate mass media.

TV Made to Order

Take a look at some of the most popular television programs in our culture: *American Idol, Survivor,* and the game shows. These are designed to reinforce the social Darwinian concept of competition. In order to survive, you must to outdo your competitors. This is the recurrent theme in our society. We have all been propagandized to have low self esteem.

The only way to feel good about oneself is to succeed. Success involves acquiring anything that is bigger and better. To succeed today, you must embellish the material things around you, particularly if you want to become someone important. That there is a difference between the individual and his image is human nature.

We are expected to accept this as a law of nature. After all, it is *survival of the fittest*. This is the way we are told that species evolve. You don't want to hold humanity back now, do you?

Think of the mass media that prays on our insecurities. Advertising, in many ways, is a con game, too: fitness magazines, fashion, obsession with celebrity, youth, wealth and beauty. We are obsessed with youth and enhancement.

Advertising is a process of manufacturing glamour. The state of being envied is what constitutes glamour. Advertising, then, is about the solitary happiness that comes from being envied by others. But envy has a dark side, which largely has been lost to the twentieth century thought.

Since medieval times, envy has been considered a major term for identifying the causes of human suffering. Like despair, envy derives from the separation of the person from the object of desire, combined with a sense that one is powerless to attain what is desired. In envy, the urge to reach out becomes the urge to destroy.

Advertising is the consumer culture's version of mythology. No society exists without some form of myth. Thus, it is not very surprising that a society which is based on the economy of mass production and mass consumption will evolve its own myth in the form of a commercial. Like myth, it touches upon every facet of life, and as a myth it makes use of the "fabulous" in its application to the mundane.

No matter how you look at it, people do not need new automobiles every third year, plasma televisions bring little enrichment of the human experience, and a higher or lower hemline offer no expansion of consciousness, no increase in the capacity to love.

Because products do not provide the kind of psychic payoff promised by the imagery of advertising, we are left to doubt whether anything can. If we follow this doubt, we wind up contemplating a state of mind in which a black hole surrounds almost every product like a ghostly negative of its radiance – the black hole of failed promise.

And into this black hole, brought about by advertising's exploitation of so many ideal images, steps any religion that promises to cut through the cycle of idolatry and connect us with the one great "Ideal" that transcends all others: God, immortality, cosmic consciousness, enlightenment, the spirit world, the deep self or whatever name *It* has.

Enter the world of virtually reality where your every dream, that "Ideal," can come true.

Virtual reality, we are told, creates an artificial world in which we are free to do whatever we want without being accountable for our actions. If you want to have sex with a computer generated image of your neighbor's dog, that's perfectly all right, go ahead and have it in the virtual world.

With that liberation of our "repressed fantasies," if one is to believe Freudian-speak: Virtual reality produces a new consciousness that redefines good and evil. In the words of that other famous psychopath, Friderich Nietzsche, it is a universe: "beyond good and evil." Thus we are liberated to do evil, "without suffering the effect of evil and therefore purge ourselves of our innate desire to 'be' evil."[34]

This fantasized pinnacle of the transhumanist world provides us with the clearest view of its deadly ends, a world seen through eyes lacking universal truth and a sacred responsibility of the sovereign individual to act for Good.

> And without universal truth, there can be no reason. If truth is killed, then our civilization is killed with it.[35]

The only change over the past fifty years has been that as technical sophistication has increased, so has circumspection. It should be of little surprise then that the advertising medium has even been reshaping our very concept of truth.

In *Brave New World*, Aldous Huxley posited the universal application of a drug called Soma for the manipulation of the populace. Now, the ultimate form of Soma enhancement is at our fingertips – transhumanism.

Transhumanism offers the ultimate form of bettering yourself. You want to be young? How does eternal life sound? You want to be strong? How about adding some machine parts to your body? You want to be smart? Here, put this computer chip in your head. You could become omnipotent and, to those who could afford it, immortal.

Genetic Treatments to Prevent the Effects of Ageing Developments in genetics:

> ... will allow treatment of the symptoms of aging and this would result in greatly increased life expectancy for those who could afford it. The divide between those that could afford to 'buy longevity' and those that could not, could aggravate perceived global inequality.[36]

Dictatorial or despotic rulers could potentially also "buy longevity," prolonging their regimes and international security risks.

If you think that transhumanism is being introduced into such a fiercely competitive world by chance, think again. We have all been trained like Pavlovian dogs with a nonstop barrage of propaganda. If we don't own up to this reality, we will all be outdone by our superiors, those people who have always believed themselves to be the pinnacle of human greatness.

And this is exactly what the National Science Foundation NBIC report postulates:

> Enhancing human performance would require merging human biology with technology.

Brain machine interfaces, would allow the control of machinery with the brain itself. Implantable brain chips would also be able to store information and enhance cognitive function. The ultimate human machine symbiosis would be to download the actual copy of a person's brain into a super computer.

I repeat, this would allow someone to effectively life forever in a computer-generated virtual simulation. And of course, the military implications of convergence are quite obvious. A cybernetic enhancement of human performance is inevitable. Achieving these visions requires the decoding and understanding of complex systems. The most important complex system being the human brain. After all, it is the driving force behind human performance.

MK-ULTRA: Back to the Future

In 1952, a proposal was issued to the Director of Central Intelligence outlining funding mechanism for highly sensitive CIA research and development project that would study the use of biological and chemical materials such as LSD, in altering human behavior.

On April 13, 1953, MK-ULTRA, was established for the express purpose of researching and developing chemical, biological, and radiological materials to be used in clandestine operations and capable of controlling or modifying human behavior.

All in all, there were 149 MK-ULTRA sub-projects, many of them involved with research into behavior modification, hypnosis, drug effects, psychotherapy, truth serums, pathogens and toxins in human tissues.

For intelligence purposes, what was required of MK-ULTRA was the ability to manipulate memory. Thus, what the Freudians call the "superego" had to be bypassed, allowing the controller direct access to the contents of an enemy agent's mind. That was step one. Step two would involve erasing specific pieces of information from the subject's memory and replacing those pieces with new bits of memory, thus permitting the Agency to send that agent back into the field without any knowledge that he or she had been interrogated and had given up sensitive information. Step three was a potential bonus: Could that enemy agent then "be programmed" to commit acts on behalf of the Agency, without knowing who gave the command or why?

This was the essence of the Manchurian Candidate. It is also the essence of what we know today as hypnotherapy and "depth" psychoanalysis, for the psychiatrist is looking for access to the patient's unconscious layers, to extract important information, such as childhood trauma, and to neutralize the effects of that trauma, in some cases replacing certain behavior patterns with new, approved patterns.

But in exploring the mind and developing techniques for unlocking its secrets, the CIA unknowingly tread on areas that have been the domain of religion and mysticism for thousands of years.

Scouring the records of occultists, magicians, witches, voodoo priests and Siberian shamans to isolate the techniques that were used since time immemorial to supplant a person's normal, comfortable, everyday consciousness, and replace it with a powerful, all-knowing and sometimes violent, and always deceptive alter personality. And to then use those alters to uncover the action of deep memory, for MK-ULTRA was, at its core, an assault on the Land of Memory; the creation of new, false memories and the eradication of old, dangerous ones.

As these monstrous notions of mass social engineering were being presented as a "humanistic" alternative to World War in the age of the atomic and hydrogen bomb, the world's intelligence agencies and governments were already hard at work on crucial projects that a generation or two later would shape the implementation of a "Brave New World," and bring us, hoi polloi, kicking and screaming to "their" new reality of Planetary Governance.

Neuroscience

Neuroscience is the study of the nervous system. With advances in chemistry, computer science, engineering, medicine and other disciplines, neuroscience now also includes the study of the molecular, cellular, developmental, structural, functional, evolutionary, computational, and medical aspects of the nervous system. From molecular and cellular studies of individual nerve cells to imaging of sensory and motor tasks in the brain, neuroscience has crossed the threshold of science and has become a key element of national security apparatus the world over.

As the power to decode the brain accelerates, changes will sweep quickly across every society in a slightly different way.

Intelligence agencies have been spending millions of dollars on neuroscience research in hopes that it will soon deliver extremely advanced tools for protecting national security. *The scope of the technology is mind boggling and a lot more advanced than governments would have you know.*

President Obama has just green lighted a ten-year 'Manhattan Project' to map the human brain. The new billion-dollar enterprise is called 'BAM,' the Brain Activity Map project. The fact that DARPA is one of the agencies involved tells you a great deal. This is the technocrats' wet dream: We can now power up an artificial brain of such power and intelligence that it can make all crucial decisions for the human race from Central Planning.

BAM will be used to devise a Minority Report society. The sub-text of BAM is: People, we will protect you. We will gain access to the brain and re-channel it away from violence. You will be safe. The stated goal is the mapping of all neuronal activity in humans. The American Psychiatric Association and Big Parma are drooling with anticipation. This is precisely what they've been aiming for. Vastly expanded treatment options. More diagnoses. More drugs. More control. BAM is the step before reconfiguring the brain, which is the long-term and underlying goal.[37]

The development of sophisticated neuro-weapons will create a perpetual state of uncertainty with the promise and peril of the development of neuro-warfare and its effects. Emotional detection systems will pervade public areas, as global surveillance networks could seek out terrorists and criminals.

The preponderance of the research now creating our emergent neuro-society is being underwritten by America's defense spending. According to Association of American Universities, nearly 350 colleges and universities hold Pentagon research contracts, amongst them Cal Tech, the University of California, California Institute of Technology, Cornell, MIT, Harvard, Yale, Princeton, Indiana University, The University of Wisconsin, University of Michigan, Penn State University, University of Minnesota, Stanford, University of Texas, University of Washington. These schools represent 60 percent of basic research funding.

President Obama's budget request for Fiscal Year 2013 included $140.820 billion for research and development (R&D). Seven federal agencies receive 95.8% of total federal R&D funding, with the Department of Defense (50.6%) and the Department of Health and Human Services (22.3% primarily for the National Institutes of Health) accounting for nearly three-fourths of all R&D funding.[38]

In an article published a decade ago by the US Army War College, military analyst, Timothy Thomas used the title: "The Mind Has No Firewall." The article examined energy based weapons, psychotropic weapons, and other developments designed to alter the ability of the human body to process the stimuli.

Humanity stands on the brink of psychotronic warfare, with the mind and the body as the focus. Psychotronic weapons are those that can remove and replace memories in the human brain. A former major in the Russian army reported in a February 1997 military journal that many weapons fitting the psychotronic definition are being developed throughout the world.

Research shows that there may have already been deployment of USSR-USA psychotronic warfare even before "The Mind Has No Firewall" was published. Disturbingly titled "Project Pandora," the research was run by the psychology division within the psychiatry research section of Walter Reed Army Institute of Research. Pandora was initiated after the United States learned that the Soviet government had from the 1953 to 1976, beamed microwave radiation at the United States Moscow Embassy.

DARPA

January 2008 issue of *Aviation Week* carried an article by Amy Kruse, a woman who oversees some of the most provocative DARPA research, such as research into computer analysis of brain waves detected from a satellite without the subject's knowledge. Which is expected to help intelligence analysts precisely identify and locate targets based on the hostile thoughts of enemy forces.

Furthermore,

> The Pentagon's Defense Advanced Research Projects Agency has tapped Northrop Grumman to develop binoculars that will tap the subconscious mind. The Cognitive Technology Threat Warning System program, informally called 'Luke's Binoculars,' combines advanced optics with electro-encephalogram electrodes that can, DARPA believes, be used to alert the wearer to a threat before the conscious mind has processed the information.[39]

What's more, the Huffington Post reports:

> The Defense Advanced Research Projects Agency has cursed the earth with unmanned missile systems, all-terrain robots and machines that feast upon, and then fuel themselves with, human flesh. The group, which works directly for the U.S. Department of Defense, now hopes to turn actual humans into controllable, mindless and murderous cyborgs.
>
> The organization has decided to further the dehumanization movement by funding an ambiguous, yet monstrous, 'transcranial pulsed ultrasound' project. According to Popular Science, the technology involves implants that manipulate troops' minds to stimulate neural processes, relieve stress, heighten alertness and mental acuity, and even reduce the effects of a traumatic brain injury. Invulnerable, hyper-sensitive warriors, who can distribute death and destruction despite irreversible brain damage, don't seem petrifying or horrific at all.[40]

In the combining interdisciplinary sciences, with our brain as a focal point, there have been experiments, where soldiers' brainwaves can seen by their commanders via wireless computers.[41] When the commander knows that the soldier has gone into tunnel vision from information overload, he will know that he will have to count on someone else for key actions and commands during an attack.

An earlier project, called "augmented cognition," gave birth to phases one, two and three of a DARPA super secret research project: neurotechnology for intelligence analysts (NIA). According to DARPA:

> Cognitive Technology Threat Warning System (CT2WS) war fighters need to be able to see and identify threats at as great a distance as possible. Binoculars have not yet integrated the technology or biology that could help maximize this capability. The Cognitive Technology Threat Warning System program will bring these technologies to develop soldier-portable visual threat detection devices. These systems will provide greater visual information about a warfighter's

surroundings while providing tools to initiate an early response when threats emerge. This program will integrate areas of technology such as flat-field, wide-angle optics, large pixel-count digital imaging, and cognitive visual processing algorithms. Other features include ultra low-power analog/digital hybrid signal processing, operator neural signature detection processing, and operator interface systems. Success from this effort will result in a composite software/human-in-the-loop system capable of high-fidelity detection with extremely low false alarm rates without adding to already significant warfighter combat loads.[42]

Phase two for the NIA project revolves around an ambitious Neovision2 program that:

> ... will develop an unattended, standalone system that can recognize relevant military objects in a wide range of ambient and environmental conditions through fusion of neuroscience and engineering. Integration of recent developments in understanding the mammalian visual pathway and advances in microelectronics will lead to the production of new revolutionary capabilities that will provide a new level of situational awareness for war fighters.[43]

The bottom line is that the technology is light years ahead of anything most of us can fathom.

One of the key corporations working on this new era technology is Honeywell, on whose payroll was Egidio Giuliani. He was a full-time terrorist operative in the 1970s and 80s, who supplied weapons and passports to both Red Brigades who killed Aldo Moro, and the "black" fascists who bombed Bologna train station.

Honeywell's work is augmented with participation from Teledyne Scientific & Imaging and Columbia University.

Teledyne is not your run-of-the-mill high tech company working on new generation washing machines. Their key projects consist of such things as:

> ... design, development and production of high performance infrared and visible sensor subsystems used in space missions, long range terrestrial surveillance and targeting

and astronomy applications and energy harvesting technologies, electronic device packaging, biomaterials, and liquid crystal-based optical devices.[44]

How far out into space is that?

In 'Minority Report' Tom Cruise's character, John Anderton, has a radical surgery to replace his eyes so that he can get past security systems that scan his retina to identify him. As he's lying in a tub recovering from his black-market procedure, tiny robots sneak into the room and scan his eyes in an attempt to track down the fugitive Future Crime officer. The ability to scan retinas to identify people is straight out of a sci-fi film, but, outside of the use of spider-like drone bots, this is very much present and near future terms. In fact, soon your eyes may not even need to be in close proximity to the scanner to be identified.

Engineers at Southern Methodist University (SMU) are working closely with DARPA to develop a new type of eye scanner that could identify a room full of people without their knowledge. The new image sensors, called Panoptes, could locate and scan a person's iris regardless of distance, and even if they're not looking directly at the camera. The system, dubbed Smart-Iris, is impervious to problems like poor lighting, glare, eye lashes, or movement. And, with the help of a new algorithm, it can function with only a partial scan.[45]

These technologies are not being developed to stop the terrorists, but rather to stop you! The laws to justify these technologies are not written on a whim. They are specifically designed to give the government carte blanche authority over the people during the chaos and confusion of the Age of Transitions. *Transition to a planetary civilization.*

You see, the future bin Ladens and Kaddafis are not the enemy. In fact, they never were. You are the enemy. Whether at the airports, border crossing or on the street corner, from now on we will be mind-probed by amazing new technology being developed by the Human Factors Division of the DHS Science & Technology directorate.

The DHS chaps themselves are relatively cagey about exactly how their mind-reader/lie-detector gear would work. The idea seems to be to employ a battery of technologies. Cameras would snap pictures or video of people's faces, which could then be automatically analyzed for suspicious expressions or memes – perhaps an anticipatory evil leer or vacant mindless drool at the prospect of finally attacking the hated Great Satan.[46]

Memetics would allow for deeper understanding of cognitive processes throughout society. The applications, of course, are oriented towards "Universal Darwinism." And as the National Science Foundation report openly states: "Certain ideas may have the force of a social virus."[47] Not surprisingly, the report also gives a visionary solution to the war on terror:

> Socio-tech can help us win the war on terrorism. It can help us to understand the motivations of the terrorists and so eliminate them.

Phase three is expected to produce a prototype that intelligence agencies will road test in 2014. According to Honeywell, the technology is nearly ready for operational use.

All together, Augmented Cognition and NIA research has linked several corporate and academic teams with the four different US military services: Daimler Chrysler with United States Marine Corp, Lockheed Martin with United States Navy, Boeing with the Air Force and Honeywell with a team of eleven industry and university partners with the Army.

One of the most talked about areas of research, is something called the "Active Denial System," ADS. ADS is a by-product of larger on-going research looking for technology that could delete, and then replace a person's memories via the use of electromagnetic radiation.

If you are thinking "Men in Black" Hollywood science fiction, you are absolutely right. Except the technology, called "Amnesia Beam" is here; ready to be used at a moment's notice. What's more, a team of neuroscientists has actually developed a brain scan based on finding hints about what a subject is intending to do. This is a nightmare version of *Minority Report* made reality.

Scientists claim that:

> The seeds of criminal and anti-social behavior can be found in children as young as three. More researchers believe that violent tendencies have a biological basis, and that tests and brain imaging can pick them up in children.
>
> By predicting which children have the potential to be trouble, treatments could be introduced to keep them on the straight and narrow. If the tests are accurate enough then a form of screening could be introduced in the same way we test for some diseases. The theories were put forward by two leading criminologists at the American Association for the Advancement of Science in Washington.[48]

Please understand, these are not publicly funded projects for the betterment of humanity, but they are mostly secret experiments sanctioned in the name of defense, which when put on its head, is crime prevention and extrapolated into the future is tailor-made to put down any rebellion by the 99% of the world's population destined to live in abject poverty in crime infested Mega-Cities of the future.

How much does all of this cost? Even though the actual numbers do not exist, what is known is that Pentagon's black operations budget is estimated at $6 billion per year.

What every one must understand is that the future is not about nuclear weapons. We are now in an arms race to create the next generation of unimaginably potent and terrifying weapons.

One such project is Honeywell's Advanced Image Triage System. It includes group behavioral analysis software and individual emotion recognition algorithms that key off micro-emotional tics we all exhibit. Intelligence analysts will scour real time surveillance feeds from satellites, unmanned predators, robotic insects and other ingenious surveillance systems in their efforts to seek and destroy enemy combatants.

The rise of neuro-weapons is here: aimed at shifting the emotional and cognitive capacity of individuals. Memory bombs that give individuals short term amnesia or electronic sleep inducing weapons may seem in the realm of science fiction, but before the advent of atomic weapons nobody imagined that a bomb would instantly kill 140,000 inhabitants of Hiroshima.

Total Spectrum Dominance

As more and more biology and chemistry labs focus on developing next generation brain drugs, other researchers working on a different piece of the neuro-technology revolution, are designing implantable medical devices that interact with the brain through tiny electrical impulses.

Over the next two decades, the impact of neuro-devices on a nano-scale will be profound. The reason so much more is at stake is that the:

> ... newest tools will give us nothing less than an increasingly precise control over the most powerful factor in our lives – our own minds.[49]

Of all the storage mediums you use to keep information that is most important to us, our brain is by far the most complex. Recent advances in brain imaging, hierarchical recurrent temporal memory, and complex brain network theory, as well as neuro-robotics are making hacking the human brain a distinct probability.

This is one of DARPA's high end reverse-engineering projects:

> With 50 million neurons (processing elements) and several hundred kilometers of axons (wires) terminating in almost on trillion synapses (connections) for every cubic centimeter, and consuming only about 12 watts energy for the entire cortex, the brain is arguably one of the most complex and densely packed, yet highly efficient information processing systems known. It is also the seat of sensory perception, motor coordination, memory, and creativity - in short, what makes us humans to humans.[50]

For now, it is being sold as the ultimate human empowerment: Your mind controlling the environment around you, as well as, the general flow of information while the sensors in your intelligent home read your subconscious mind's secret desires. Via implants you are in a constant mental contact with your loved ones.

> Everything and everyone is included in the Great Cloud where man and machine, and everything else in the cosmos eventually form one big, harmonious happy family without the presence of irrational terrorism and irrational violence or other uncertainties. Heaven on earth realized with the help of technology.
> The danger is that the new technology allows a total control of citizens, not only regarding where they are and what they do, but also what they think about and intent to do – as soon as the thought emerges into their minds for the first time. Your secret, erotic fantasies and all your passwords and so on, become impossible to hide from the person sitting in the other side of the line of the equipment for mind reading. The concept of 'big brother is watching you' gains a whole new dimension.[51]

The European Union is heavily involved in this research. One of their key initiatives is "Project Quasar" whose objective:

> … is to pull together all our existing knowledge about the human brain and to reconstruct the brain, piece by piece, in supercomputer-based models and simulations.[52]

Project Quasar is part of the Human Brain Project (HBP), one of the EU's two Future and Emerging Technologies (FET) initiative Flagship directives.

> The project aims to combine information about users to build business models that provide a more efficient use of the available frequency spectrum.[53]

EU wants to connect the dots leading from genes, molecules and cells to human cognition and behavior. Buried in the report and never mentioned in public is a framework dealing with the problem of how to fuse humans and machines.

In 2011, a team of scientists created a chip that can control the brain and can be used as a storage device for long-term memories.

> In studies the scientists have been able to record, download and transfer memories into other hosts with the same chip implanted.[54]

This may sound like an amazing break through, but what few realize is that the advancement in technology brings the world one step closer to a global police state and the reality of absolute mind control.

> More terrifying is the potential for implementation of what was only a science fiction fantasy – the 'Thought Police' – where the government reads people's memories and thoughts and can then rehabilitate them through torture before they ever even commit a crime based on a statistical computer analysis showing people with certain types of thoughts are likely to commit a certain type of crime in the future.[55]

Too far-fetched for your liking? We have already preemptively invaded nations and tortured alleged terrorist suspects with absolutely no due process of law, so the idea of preemptively torturing innocent people is just another step in the destruction of our civil liberties.

> Perhaps a less sensational example is depicted in the modern day *Matrix* movies, in which computer programs are uploaded into people's brains allowing them to instantly learn how to perform a wide variety of tasks.[56]

The Ties that Bind

What has been very much under the radar is the close, long-term relationship that exists between the US military and Big Pharma to discover new pharmacological and training approaches that will lead to an extension of the individual warfighter's cognitive performance capability.

Back in 1999, if a male in the USA had a history of taking psychiatric drugs, even including Ritalin, this could be an automatic disqualification from enlisting in the military. Now, a mere

14 years later, soldiers are given packs of these drugs to take on the battlefield, including antidepressants, and they self-medicate.

Among them, cognitive-augmentation drugs such as modafinil, which enhances alertness even after long hours of wakefulness, are becoming widespread. That's a huge paradigm change, and it has to do with a deal that was cooked up between the US military and Big Pharma.

However, the self-medicating soldiers still are involved in making certain personal choices. One of these is whether to take the drugs or not. The military would to eliminate that pesky free will and the element of the unknown. How?

How about micro chipping your brain?

> That might not sound very appealing to you at this point, but this is exactly what the big pharmaceutical companies like GlaxoSmithKline and the big technology companies have planned for our future.[57]

Again, the key word is *control*, in all of it manifestations. The National Science Foundation NBIC 2001 report spends an exorbitant amount of time on examining control and human behavior. I quote:

> The multiple drivers of human behavior have long been known. Now, through the decoding of complex systems a completely predictable and managed society can be realized.... To use the tremendous computing power we now have to integrate data across those fields to create new models and hence new understanding of the behavior of the individuals. The ultimate goal is acquiring the ability to predict the behavior of an individual, and by extension, of the group ... using tools and approaches provided by science and technology – will raise our ability to predict behavior. It will allow us to interdict undesirable behaviors before they cause significant harm to others and to support and encourage behaviors leading to greater social good.[58]

Millions of dollars are being pumped into researching cutting edge technologies that will enable implantable microchips to greatly "enhance" our health and our lives. Of course nobody is going to force you to have a microchip implanted into your brain, when they are first introduced.

As the *Financial Times* reports:

> Diseases such as diabetes and epilepsy and conditions such as obesity and depression will be treated through electronic implants into the brain by sending electrical signals to malfunctioning cells rather than pills, injections or surgery.[59]

And as *Financial Times* asks:

> If a brain implant could cure a disease that you have been suffering from your whole life would you take it?

The Journey Toward Making "Normal" Obsolete

One of the ways of manipulating human brain activity is through optogenetics, a revolutionary new form of wireless communication in which nerve cells in the brain are programmed genetically, which allows you to remote control your brain activity with light.

> The technique involves the genetic modification of neurons to make them produce opsins – light-sensitive proteins that are normally made in photoreceptor cells in the eye and in some micro-organisms. These ontogenetic neurons can then be switched on and off with different light signals, making brain activity controllable.

Financial Times reports:

> Human optogenetics will require two operations. First the light-sensitive genes are introduced to the patient's neurons – probably carried in a harmless virus, as with other applications of gene therapy. Later a fine fiber-optic cable is inserted through a small hole drilled in the skull to illuminate the target area of the brain.[60]

When the "benefits" of such technology are demonstrated to the general public, soon most people will want to become "super-abled" No longer will there be a need to take harmful, anabolic steroids to run faster, jump higher and endure longer.

What was once science fiction is rapidly becoming reality, and it is going to change the world forever.

Too good to be true, right? According to an article in the *Wall Street Journal*, the typical procedure is very quick and it often only requires just an overnight stay in the hospital:

> *Neural implants, also called brain implants, are medical devices designed to be placed under the skull, on the surface of the brain. Often as small as an aspirin, implants use thin metal electrodes to 'listen' to brain activity and in some cases to stimulate activity in the brain. If that prospect makes you queasy, you may be surprised to learn that the installation of a neural implant is relatively simple and fast. Under anesthesia, an incision is made in the scalp, a hole is drilled in the skull, and the device is placed on the surface of the brain.*[61]

So, *corporate mass media is again being the principal booster and mind bender of the general public's perception of reality. R*ight off the bat, corporate media is telling us how to interpret neural implants. Sounds like a "High Tech for Dummies 'How To' guide."

You are being told what to think and how to interpret the information. Of course, in order to interpret it "correctly" you must have the right mindset. And, in order to get the unsuspecting public into the right mindset for the government's sinister plans, they need the official blessing of the leading mainstream corporate media publications. How?

For example, the *Wall Street Journal* reports that:

> *These tools aren't sinister. They're being created to solve real problems. Simply put, prosthetic limbs help people move, and neural implants help people think…. The technology can give us brains and brawn. All we have to do is let the devices under our skin.*

One of the poster boys for the entire Transhumanist movement is a now fallen angel, South African athlete, Oscar Pistorius, who happens to have had both of his legs amputated below the knee. The mind-benders and paradigm-shift planners used the London

Olympics, the world's greatest sport's venue to showcase their big idea. Using upside down question mark-shaped carbon fiber sprinting prosthetics, called Cheetah blades, Mr. Pistorius, became one of the most recognized athletes in the world. This was not an accident.

Buried deep in the *Wall Street Journal* article is your classic definition of eugenics and selection of the fittest and the richest over the oppressed masses of the great unwashed.

> The dissemination of advanced implantable technology will likely be just as ruthlessly democratic as the ailments it is destined to treat. Meaning that, someday soon, we may have a new class of very smart, very fast people– yesterday's disabled and elderly. That's right, yesterday's disabled and elderly, because they will have all been eliminated as the prohibitive costs and/or the natural right to be given the implants, will have precluded these poor wretches from having access to this technology.[62]

Think of people wanting to upgrade themselves like the latest model of your cell phone. Those who don't or who can't will be left hopelessly behind.

You see, there is a lot of money to be made here through legitimate channels and on the black markets. By the way, both channels will be controlled by the same financial-political apparatus. Just as the international drug trade and distribution is controlled by the banking and the military-industrial cartel.

Control will be exercised through technology. Integrating technology with human evolution sounds like such a good idea, at first. "You are building a better you. It is a business that's driven by fear. If I don't improve myself.

Augmentation can be a huge business for these corporations.

> We have been integrating ourselves with technology for decades now, replacing damaged limbs with mechanical limbs, implanting data chips into our bodies which give away huge amounts of information to governments and corporations all across the world. Has it come to the point when we will be actively encouraged to exchange our perfectly functional body parts for upgraded applications?

Is it a question of ethics? Morality? Again, are we/they playing God with the human race? And at what cost? What will Planet Earth look like in one generation?[63]

When you fuse human augmentation with control over it, you get military grade augmentation. Cyborgs, the future of our planet?

> But, it comes at a price. You will have to take the drugs for the rest of your lives to make sure the augmentation works. These drugs are both dangerous and addictive and expensive. If you don't, your body will reject your augmentation. The elite will have their technology in you. They have the power to turn off your limbs, the potential to turn off your eyes, send messages to your brain and control your thoughts as if they have the power of God.[64]

This is a top-down system created by the elite to benefit a very few. We are at the bottom, but if we all wake up and realize that this is not *our* system, only then can it be changed.

Human experiments? What is immortality? What does it mean to be immortal? Where does it end? The end of the human species?

But technology doesn't end there:

> Intel is working on sensors that will be implanted in the brain that will be able to directly control computers and cell phones. *By the year 2020, you won't need a keyboard and mouse to control your computer. Instead, users will open documents and surf the web using nothing more than their brain waves. The brain waves would be harnessed with Intel-developed sensors implanted in people's brains.*[65]

The potential "benefits" of such technology are almost beyond imagination. An article on the website of the *Science Channel* put it this way:

> If you could pump data directly into your gray matter at, say, 50 mbps – the top speed offered by one major U.S. internet service provider – you'd be able to read a 500-page book in just under two-tenths of a second.[66]

How would the world change if you could download a lifetime of learning directly into your brain in a matter of weeks?

Man Digital 2.0

The possibilities are endless. But so is the potential for abuse. Through interaction, implantable microchips can "talk" directly to the brain, bypassing the sensory receptors. This would give a tyrannical government an ultimate form of control.

> If you could download thoughts and feelings directly into the brains of your citizens, you could achieve total control and never have to worry that they would turn on you.[67]

But, we haven't even scratched the surface. Called Remote Neural Monitoring, RNM, the technology already in use in the USA, UK, Spain, Sweden, Germany and France, allows them to see through your eyes, hear your thoughts, and upload photos and scents into your brain as real as if you saw or smelled it in the natural environment.

Needless to say, the perpetrators can hear what you hear because you become a unit of the mainframe. They can change your behavior, affect memory functions and emotions. This is not a plotline of a dystopian novel. This is real and it is being implemented today, every day by the governments who profess to protect us from evil.

In fact, you could potentially program these chips to make your citizens feel good all the time: The ultimate goal of Huxley's scientific dictatorship without tears – soma personified. The future is now.

Instead of drugs like cocaine and marijuana giving you a natural high, or shooting you to the Moon, you could have these chips produce a "natural high" that never ends. Drug dependency replaced by a fully government sanctioned chip dependency. *The way of the future.*

> RNM has a set of certain programs functioning at different levels, like the signals intelligence system which uses electromagnetic frequencies (EMF), to stimulate the brain for RNM and the electronic brain link (EBL). The EMF Brain Stimulation system has been designed as radiation intelligence, which means receiving information from inadvertently, originated electromagnetic waves in the en-

vironment. However, it is not related to radioactivity or nuclear detonation. The recording machines in the signals intelligence system have electronic equipment that investigate electrical activity in humans from a distance. This computer-generated brain mapping can constantly monitor all electrical activities in the brain. The recording aid system decodes individual brain maps for security purposes.

For purposes of electronic evaluation, electrical activity in the speech center of the brain can be translated in to the subject's verbal thoughts. RNM can send encoded signals to the auditory cortex of the brain directly bypassing the ear. This encoding helps in detecting audio communication. It can also perform electrical mapping of the brain's activity from the visual center of the brain, which it does by bypassing the eyes and optic nerves, thus projecting images from the subject's brain onto a video monitor. With this visual and audio memory, both can be visualized and analyzed. This system can, remotely and non-invasively, detect information by digitally decoding the evoked potentials in 30-50Hz, 5 millwatt electromagnetic emissions from the brain. The nerves produce a shifting electrical pattern with a shifting magnetic flux that then puts on a constant amount of electromagnetic waves. There are spikes and patterns that are called evoked potentials in the electromagnetic emission from the brain. The interesting part about this is that the entire exercise is carried out without any physical contact with the subject.

The EMF emissions from the brain can be decoded into current thoughts, images and sounds in the subject's brain. It sends complicated codes and electromagnetic pulse signals to activate evoked potentials inside the brain, thus generating sounds and visual images in the neural circuits. With its speech, auditory and visual communication systems, RNM allows for a complete audio-visual brain to brain link or a brain-to-computer link.

Of course, the mechanism needs to decode the resonance frequency of each specific site to modulate the insertion of information in that specific location of the brain. RNM can also detect hearing via electromagnetic microwaves, and it also features the transmission of specific commands into the subconscious, producing visual disturbances, visual hallucinations and injection of words and numbers in to the brain through electromagnetic radiation waves. Also, it manipulates emotions and thoughts and reads thoughts remotely, causes pain to any nerve of the body, allows for remote manipulation of behavior, controls sleep patterns through which control over communication is made easy.[68]

Is any of this being implemented today? Indeed, it is.

THOUGHT CRIME POLICE

John St. Clair Akwei, a former National Security Agency employee and whistle-blower exposed the use of neural monitoring to spy on individuals.[69] This is a key area involving mind-invasive technologies that are playing a role in the uncovering and punishment of "Thought Crime."

NSA, the National Security Agency is America largest and most powerful organization.

> The Signals Intelligence mission of the NSA has evolved into a program of decoding EMF waves in the environment for wirelessly tapping into computers and tracking persons with the electrical currents in their bodies.
>
> How is it being done? It's done by EMF or ELF Radio Waves, and a technology known as 'Remote Neural Monitoring'. Signals Intelligence is based on the fact that everything in the environment with an electric current in it has a magnetic flux around it that gives off EMF waves. Both the NSA and the Department of Defense has developed proprietary advanced digital equipment that can remotely analyze all objects whether man-made or organic that has electrical activity.
>
> NSA Signals Intelligence uses EMF Brain Stimulation for Remote Neural Monitoring (RNM) and Electronic Brain Link (EBL). EMF Brain Stimulation has been in development since the MKUltra program of the early 1950s, which included neurological research into radiation (non-ionizing EMF) and bioelectric research and development. The resulting secret technology is categorized at

the National Security Archives as 'Radiation Intelligence,' defined as 'information from unintentionally emanated electromagnetic waves in the environment, not including radioactivity or nuclear detonation.'

NSA computer-generated brain mapping can continuously monitor all the electrical activity in die brain continuously. The NSA records and decodes individual brain maps (of hundreds of thousands of persons) for national security purposes. EMF Brain Stimulation is also secretly used by the military for Brain-to-computer link. (In military fighter aircraft, for example.) For electronic surveillance purposes electrical activity in the speech center of the brain can be translated into the subject's verbal thoughts. RNM can send encoded signals to the brain's auditory cortex thus allowing audio communication direct to the brain (bypassing the ears). NSA operatives can use this to covertly debilitate subjects by simulating auditory hallucinations characteristic of paranoid schizophrenia. Without any contact with the subject, Remote Neural Monitoring can map out electrical activity from the visual cortex of a subject's brain and show images from the subject's brain on a video monitor. NSA operatives see what the surveillance subject's eyes are seeing. Visual memory can also be seen. RNM can send images direct to the visual cortex bypassing the eyes and optic nerves. NSA operatives can use this to surreptitiously put images in a surveillance subject's brain while they are in R.E.M. sleep for brain-programming purposes.[70]

Psychotropic Weapons

Achievements in science, engineering, medicine, genetics, digital information and communication technologies have made possible today's terrifying component psychophysical weapon (CPF). CPF to the XXI century are what atomic weapons were to the XX century. The effect of these weapon systems is far more deadly, albeit invisible to the naked eye. Mind control via non-lethal weapons.

The psychotropic weapon at its most diabolical is a directed energy system installed in outer space that can influence the brain and the central nervous system: changing the mental, emotional and behavior functions of the person.

The majority of them operate in a wave band of a brain, modifying consciousness, a physical and mental condition of the person. Through psychotropic weapons, behavior of the entire world's population or specific regions of the earth can be controlled in a precise and deliberate way.

Neurophysicists have discovered that they can model the vibratory patterns of psychotropic chemicals like LSD. When scientists project these carefully shaped sonic vibrations onto the human brain, they can induce altered states of mind, exactly as if they were injecting the person with a drug. An imbalance occurs, a fundamental change in a person's psyche, he loses self-control and becomes easily led, and his mind moves from the real world to a world of hallucination.

The powerful hypodermic effect of such sonochemisty has some terrifying and very scary applications. It means that electro-magnetic beam weapons can be used to 'drug' people against their will. Beam weapons can be used remotely, at a distance. When combined with space-based satellite systems, such beam weapons could potentially drug entire populations, *en masse*.[71]

How is it done?

Let's start by saying that every object on this planet has an electrical frequency called Hertz (Hz). Our brain, for example, operates on electrical current. Our body is a radiant machine "in which there are complex biochemical processes with a range of frequencies from 0-100 Ggz. So bodies work in low-frequency area 0-50 Gz, the groups of cells forming fabrics, in kilogerz a range, cells 'communicate' in a millimetre wave band 40-70 Ggz. Energy information the exchange of chemical and biological and wave character inside an organism and with an environment is carried out by genetic program DNO and nervous system.

> The most universal target for psychophysical weapons is the brain of the person which adjusts functioning all systems of an organism, carries out thought processes and defines behaviour of the person. The brain works in a range of 4 kinds of 'brain waves.'

The most part of day the brain works in a range Beta waves 14-20 gz. There are three more such as 'brain waves': the Alpha wave 8-13 gz – perception of training, an easy relaxation. Theta waves 4-7 gz – ideas, images, programming of subconscious, a deep relaxation, meditation. Delta waves 0-3gz – dream, dreams, make active of immune system. Ranges of own fluctuations of a brain of the person, and also organs, substance and cells were taken into account by development psychophysical weapons.

The psychophysical weapon includes:

The psychotropic weapon – the pharmacological means of chemical and biological character having a range of own fluctuations, equivalent to rhythms of a brain and the central nervous system, and thereof influencing functioning of a brain.[72]

A Member of the Russian Federation of Space Exploration Scientific and Technical Council, Anatoliy Ptushenko, states emphatically that psychotropic weapons:

> Do not enable the individual human mind to be controlled in a precise and purposeful way. They simply 'jam' any internal connections responsible for a person's self-control, and he becomes easily controllable 'according to mob law' in line with commands form a space-based station. He can be controlled either from earth or from a command center lost in space.

How are they to be used?

I quote from a scientific paper on the subject taken from October 1996 edition of *Armeiskii Sbornik* (Moscow):

> The advanced global satellite communications at low altitude Teledesic is of special interest. It will have 15 times more satellites Iridium-840. All things being equal, the Low Earth Orbit (no more than 700 km) of small light aircraft allows the growth of their radio emissions on the Earth's surface 2,500 times or more, and allows the execution of a wide range of military missions. This is something unprecedented: the numerical size of the Iridium or-

bital grouping enables at least the simultaneous irradiation of any point on Earth from two spacecraft. This provides a double redundancy and increased reliability of communications, such as for military systems. The same emission band of radio frequencies (20-30 GHz) has never been used before in commercial communications.

An analysis of the characteristics listed indicates that the Teledesic system can be used to radiate stations located on land, at sea and in the air with emissions modulated high-power, which in a number of automated control systems used to initialize computer viruses as the so-called "dormant", which are triggered by a special signal. This can become a real threat to the security of those countries whose system of command and control are based on foreign equipment.

A psychophysical effect on people is similarly possible, in order to alter their behavior and even controlling the social aims of regional or even global society. Fantasies? But the fact is that currently the United States as matching funds are investing in the development of psychotropic weapons as the most complex space programs, and a similar correlation cannot be accidental. The Americans have started this research in the pre-war period and have continued after the war as part of programs known as MK-Ultra mind control, remote alteration of human behavior MK-Delta, as well as Bluebird and Artichoke.

So, the new space systems are potentially dangerous from the point of view of the possibility of unleashing an 'information war' on a large scale and also to create global systems for controlling the behavior of people in any region, city or locality, including one where we live. The nation that possesses these systems will win a huge advantage.[73]

The Future

Many of the transhumanists and mind benders fervently believe that in the next several decades we'll have computers into which you'll be able to upload your consciousness – the mysterious thing that makes you, you.

Spectrum Magazine editor Glenn Zorpette writes:

> The brain is nothing more, and nothing less, than a very powerful and very odd computer. Evolution has honed it over millions of years to do a fantastic job at certain things, such as pattern recognition and fine control of muscles. The brain is deterministic, meaning that its reactions and responses, including the sensations and behavior of its "owner," are determined completely by how it is stimulated and by its own internal biophysics and biochemistry. Given those facts, most mathematical philosophers conclude that all the brain's functions, including consciousness, can be re-created in a machine. It's a matter of time.[74]

Rodney Brooks, founder of MIT's Humanoid Robotics Group agrees:

> I, you, our family, friends, and dogs – we all are machines. We are really sophisticated machines made up of billions and billions of biomolecules that interact according to well-defined, though not completely known, rules deriving from physics and chemistry. The biomolecular interactions taking place inside our heads give rise to our intellect, our feelings, and our sense of self. Accepting this hypothesis opens up a remarkable possibility. If we really are machines and if--this is a big *if*--we learn the rules governing our brains, then in principle there's no reason why we shouldn't be able to replicate those rules in, say, silicon and steel. I believe our creation would exhibit genuine human-level intelligence, emotions, and even consciousness.

To bring society down to the level of a beast, it is especially important from the point of view of the elite, to control the planet Earth.

> Since the only source of increase of mankind's power, as a species, within and upon the universe, is that manifold of validated discoveries of physical principle, it follows, that the only form of human action that distinguished man from beast, is that form of action, which is identified as cognition, by means of which the act of discovery of accumulated validatable universal physical principles is generated. It is the accumulation of such knowledge

for practice, in this way, from generation to generation, which defines the provable evidence of the absolute difference between man and beast.[75]

And for that matter between man and a machine.

With everything you have read here, think about the following: How often have we chased the dream of progress only to see that dream perverted? More often than not, haven't the machines we built to improve life, shattered lives of millions, destroyed our ability to love, aspire or make moral choices – the very things that make us human? It also risks giving a few men the power to make others what they choose – regardless of the cost to human dignity.

Life extension, singularity and the promise of the golden age. Are we to believe in such a thing? Or are we being led down the proverbial road to Hell? Post human, a word not to be taken lightly, as it implies the end of the human race. Who has the authority to make such a decision? Only a God, or a power that believes itself to be a God can take such action.

Endnotes

1   Bill Clinton, New Year's Eve speech, December 31, 1999.
2   NBIC, x.
3   NBIC, p xii.
4   Sarif industries.
5   Film: *Prometheus*.
6   Aaron Dykes, "United Nations Envisions Transhumanist Future Where Man is Obsolete," Infowars.com, June 10, 2012.
7   *Russia 2045*.
8   *Russia 2045*.
9   *Avatar, A New Dawn*, Tal Brooke, http://www.scp-inc.org/publications/journals/J3402/index.php.
10   "Upload Your Brain Into A Hologram: Project Avatar 2045 - A New Era For Humanity Or Scientific Madness? July 21, 2012.
11   Aaron Dykes, United Nations Envisions Transhumanist Future Where Man is Obsolete," Infowars.com, June 10, 2012.
12   Kent Bain.
13   Servando Gonzalez, *Psychological Warfare and the New World Order*.

14   *Russia 2045.*
15   May 1974, Contract Number URH (489)-2150-Policy Research Report No. 414.74.
16   SRI was also deeply involved in the notorious MK-ULTRA program. Harman was a long time president of the Institute of Noetic Sciences and a friend to Edgar Mitchell, also of IONS, who in turn, was a long time friend of George Bush Sr. (Both are 33rd degree Scottish rite Masons).
17   Aldous Huxley, *Brave New World Revisited.*
18   Ray Kurzweil, *The Singularity is Near: When Humans Transcend Biology*, Viking Press.
19   Michael Minnicino, "Drugs, Sex, cybernetics, and the Josiah Macy Jr. Foundation," *EIR*, July 2, 1999, Vol 26, number 27.
20   "The Humbuggery of Charles Darwin, Ann Lawler, July 23-24 National Conference of the Citizens Electoral Council, November 25, 2011, *EIR.*
21   Ibid.
22   Ibid.
23   Ibid.
24   Ruth Barton, *The X Club: Science, Religion, and Social Change in Victorian England*, Philadelphia: University of Pennsylvania, 1976.
25   Ann Lawler, "The Humbuggery of Charles Darwin, July 23-24 National Conference of the Citizens Electoral Council, November 25, 2011, *EIR.*
26   Engdahl, p.77.
27   Engdahl, p.75.
28   In 1926, it awarded $250,000, an equivalent of $30 million in 2012 dollars, an astonishing sum of money in a Germany devastated by Weimer hyperinflation and economic depression.
29   Bar-Yam, Y. 1997. *Dynamics of complex systems.* Cambridge, Perseus Press.
30   *NBIC* report, page 33.
31   Lonnie Wolf, Turn off your TV, *The New Federalist*, p.87.
32   Chris H. Gray ed., *The Cyborg Handbook* (New York: Routledge, 1995. In the Soviet Union, much of the cosmonaut-training program came under the control of the Soviet Council on Cybernetics. See Slava Gerovitch, "'New Soviet Man' Inside Machine: Human Engineering, Spacecraft Design, and the Construction of Communism," in Greg Eghigian, Andreas Killen, and Christine Luenberger (eds.), *The Self as Project: Politics and Human Sciences* (Chicago: University of Chicago Press, 2007).
33   http://www.nanoweapons.co.uk/.
34   Lonnie Wolf, "Turn off your TV," *The New Federalist*, p. 84.
35   Ibid.
36   http://davidrothscum.blogspot.com/2009/03/autism-eugenics-and-split-within.html.
37   Jon Rappoport, "The hideous BAM in Obama: map your brain for your own good," February 18, 2013, www.nomorefakenews.com.

38	John F. Sargent Jr., Federal Research and Development Funding: FY2013, October 1, 2012.
39	Sharon Weinberger, "Northrop To Develop Mind-Reading Binoculars," 06.09.08.
40	"DARPA'S 'Pulsed Ultrasound' Helmets Could Control Soldiers' Minds," Warren Riddle on September 11, 2010.
41	"Can a satellite read your thoughts?" By Deep Thought, Wed Oct 24, 2012
42	http://www.darpa.mil/Our_Work/DSO/Programs/Cognitive_Technology_Threat_Warning_System_%28CT2WS%29.aspx.
43	http://www.darpa.mil/Our_Work/DSO/Programs/Neovision2.aspx.
44	http://www.teledyne-si.com/.
45	"DARPA's Smart-Iris Can Detect Eyes in a Moving Crowd," Terrence O'Brien on June 1, 2010, *Huff Post Tech*.
46	"Project Hostile Intent plans 'non-invasive' DHS brain scan," August 9, 2007.
47	NSF 2001 report.
48	"Child brain scans to pick out future criminals," Richard Alleyne, *London Telegraph*, Feb 22, 2011.
49	http://www.youtube.com/watch?v=Qu-1zO29IxE.
50	http://brain.kaist.ac.kr/about_us.html.
51	http://nanobrainimplant.wordpress.com/2013/02/11/are-people-superfluous/.
52	"The Human Brain Project Wins Top European Science Funding," News MediaCom, January 28, 2013.
53	http://nanobrainimplant.wordpress.com/2013/02/11/are-people-superfluous/.
54	"Scientists Successfully Implant Chip That Controls the Brain," *Activist Post*, June 19, 2011.
55	Ibid.
56	Ibid.
57	Michael Snyder, "They Really Do Want To Implant Microchips into Your Brain," *American Dream*, August 2, 2012.
58	NSF 2001 report.
59	Clive Cookson, "Healthcare: Into the cortex," July 31, 2012, *Financial Times*.
60	Ibid.
61	Daniel H. Wilson, "Bionic Brains and Beyond," *Wall Street Journal*, June 1, 2012.
62	Ibid.
63	Sarif industries.
64	Ibid.
65	Sharon Gaudin, "Intel: Chips in brains will control computers by 2020," November 19, 2009, *Computerworld*.

66   http://blogs.discovery.com/good_idea/2009/06/downloading-data-directly-into-your-brain.html.

67   Michael Snyder, "They Really Do Want To Implant Microchips into Your Brain," *American Dream*, August 2, 2012.

68   http://www.mindcontrol.se/?page_id=38.

69   *Akwei v. NSA* 92-0449, http://www.iahf.com/nsa/20010214.html.

70   Magnus Olsson, "Thought Police – Total Surveillance Society," *MindTech* Sweden, February, 14, 2009.

71   http://moscomeco.narod.ru/2e.htm.

72   Ibid.

73   Major General Valeriy Menshikov, Colonel Boris Rodionov, *Moscow Armeyskiy Sbornik*, Oct. 96 No. 10, pp. 88-90.

74   Glenn Zorpette, "Waiting for the Rapture," *Spectrum Magazine*, June 2008.

75   Lyndon LaRouche, "Star Wars and Littleton," June 11, 1999, *EIR*.

Epilogue

As I write these lines, we are already halfway through the year 2013. I look around me and ask myself again and again, what does it mean to be human? "Close proximity to a majestic mountain is a mixed blessing," noted Edward Said, "One is at once graced by the magnanimity of its pastures and the bounty of its slopes," and yet one can never see where one is sitting, under the shadow of what greatness, or the embracing comfort of what assurance.

Are we, as a human race, in danger of extinction? If yes, then from whom? Against the backdrop of breathtaking technological development, the urge to change oneself, to upgrade oneself, to transmogrify oneself into something better, superior, more durable, immortal ... against the backdrop of unique human fragility.

We have gone from meaningless generalizations, from the banalities and simplistic extrapolations of early science fiction novels about deep sea monsters, electric submarines, and a space cannon Moon landing to nanotechnology, robotics, cybernetics, artificial intelligence, life extension, brain enhancement, brain-to-brain interaction, virtual reality, genetic engineering, teleportation, human-machine interfaces, neuromorphic engineering.

It won't be long before we compete with God on equal terms for immortality.

Much of what I have written in this book does not have a conclusion or an ending, happy or otherwise, because, self-evidently, the future has not happened yet. What is beyond any doubt is, when we extrapolate what we know about this world, about the technology and about technological progress, we can make fairly accurate predictions of what to expect of the world.

Despite the tendency of people to confront the challenges of the future on their own terms, without the context of history and human experience, I make no such attempt in this book.

As John Gray has said, "People who worry about problems that others are not worrying about are irritating and are disparaged after the event. People who were right when others were wrong are even more irritating."

I am convinced that the future will happen as a result of long-wave themes and developments that unite the past, the present and the future. However, one constant evident in history – the power of contingency and surprise – will continue to dominate our future, which will be influenced and punctuated by unexpected events, startling surprises, major discontinuities and the pervasive operation of chance. And in all of it, we, the people, shall remain history's main protagonists.

Let me say it another way: Parts of the projected landscape are unlikely to survive first contact with the future, mainly and inconveniently because of the tendency of human beings to interfere with the scenery and to act and react in unforeseen, non-linear ways. Again, the pesky humans are getting in the way. God bless us for that!

Taken together, I believe that this work provides a complex, but readily discernible tapestry of possible outcomes, a solid map, to help you navigate the treacherous waters of uncertainty that awaits us in the very near future.

We stand at the threshold of a new renaissance in science and technology, based on a comprehensive understanding of the structure and behavior of matter from the nanoscale up to the most complex system yet discovered – the human brain.

Will our quest for immortality over-shadow the values that make us human? Will our desire to be free overcome the elites' drive for complete control? I don´t know, but it won't take us long to find out.

Daniel Estulin
May 6, 2013

Index

A

Adorno, Theodore 152, 156
Agent Orange 57
AIG 60
Akwei, John St. Clair 213, 222
Akzo Nobel 28
Alexander the Great 80
Amazon.com 94, 97, 99, 107
American Enterprise Institute 105, 160
American Eugenics Society 180
Amstutz, Dan 38
AOL-Time Warner 107
Apple, Inc. 89, 98, 104-105
Archer Daniels Midland (ADM) 43-46
Aspen Institute 105
Associated Press (AP) 111
AstraZeneca 44
Astronautics 188
Atlantic Council 105
AT&T 102
Attali, Jacques 31
Avatar (movie) 126, 163
Aviation Week 197
AXA 28

B

Baker, Kevin Robert 137, 155
Balfour, Arthur 174, 180
Ball, George 26
Balsemão, Francisco Pinto 3
Baltic Mercantile and Shipping Exchange 45
Banco Espirito Santo 19
Banco Santander 19
Bank of America 28
Banque Worms 28, 111
Barclays PLC 60
BASF Chemical Co. 44, 66
BASF Plant Science, 44
Bayer AG 59, 66
Bayer CropScience 44
Beachey, Roger 43
Bear Stearns 60
Beatrix (Queen) 3, 17
Becker, Hal 81
Bernays, Edward 82, 113
Bernhard (Prince) 183
Biddle, Mary Duke 180
Bilderberg 2-3, 11, 20-21, 26, 40, 43, 49, 74-75, 79-80, 86-87, 89, 98, 100, 103-104, 108, 110-112, 164-165, 169, 171, 183
BioRad Laboratories 135-136
Black Nobility 17, 19, 41, 80
Blue Bird, The 92
Blumenthal, W. Michael 21
Boeing Company 28, 54, 102, 145, 169, 201
Boston Globe 146
British Petroleum (BP) 28-29, 76, 78
Brave New World 9, 76, 162, 168, 183, 185, 192
Brave New World Revisited 220
Breyer, Jim 96, 103
British East India Company (BEIC) 177
British Eugenics Society 181-183
Brookings Institution 105
Brooks, Rodney 218
Brzezinski, Zbigniew 21, 49, 164
Bunge y Born 44
Burgess, Anthony 8
Bush, George H.W. 49

C

Cargill, Inc. 38, 43-46, 57
Carnegie Endowment 105, 164, 173
Carrasco, Andres 58
Carter, Jimmy 21, 49-50
Central Intelligence Agency (CIA) 93, 101-102, 114, 135, 188, 194
Chambers, Robert 179
Chancellor, John 86, 113
Chase Bank 53
Cheney, Dick 103
Churchill, Winston 50, 180
Ciba-Geigy Corp. 44
Citigroup Inc. 19, 60
Clinton, Bill 157, 165, 219
Clockwork Orange, A 8
Club of Rome 17, 41, 105
Clynes, Manfred 188

225

Codex Alimentarius 65-68, 78
Commerzbank 19
Continental Grain Company 44-46
Corporation for Public Broadcasting 111-112
Council on Foreign Relations (CFR) 3, 18-21, 49, 863, 110, 165
Coutts & Co. 19
Crowley, Aleister 183
Cruise, Tom 200

D

Darwin, Charles 2, 174-176, 179, 183-184, 220
Darwin, Erasmus 179
Davenport, Charles 180
Davignon, Etienne 3
Davis, John 59
Defense Advanced Research Projects Agency (DARPA) 103, 126-129, 145-151, 155-156, 160, 163, 196-198, 200, 203, 221
Deutsche Bank 60
Dodge, Cleveland 180
Doors of Perception 185
DowAgroSciences LLC 44
Dow Chemical Co. 44, 46, 56
Dresden Kleinwort Benson 29
DuPont Chemical Co. 44, 46, 56, 57

E

EBay 99, 107
Economist, The 29, 52, 109-111
Eli Lilly Co 44
Emery, Frederick 6, 8-10, 83, 84
Engdahl, William 34, 47, 50, 52, 76-78
Enigmas of Life 175
Essay on the Principle of Population, An 175
Eugenics Quarterly Magazine 181
European Union (EU) 33, 112, 204

F

Facebook 87-88, 91, 95-99, 114
Financial Times 29, 109, 207, 221
Flexner Report, The 61
Food and Agriculture Organization (FAO) 65-68

Food and Drug Administration (FDA) 57-58, 62, 63
Ford Foundation 105
Ford, Gerald 49
Fox News 129, 155
Frankfurt School 7, 19, 152
Freud, Sigmund 82, 88, 107

G

Galton, Francis 75, 174
Gamble, Clarence 180
Gates, Bill 96
Geithner, Timothy F. 3
General Agreement on Tariffs and Trade (GATT) 37-40, 65
General Electric 28
Gingrich, Newt 159
Giuliani, Egidio 199
GlaxoSmithKline 28-29, 206
Goldberg, Ray 59
Goldman Sachs 3, 28, 60
Google.com 88-89, 92, 94, 97-102, 104-106, 113-114
Gorbachev, Michael 17, 105
Graham, Katherine 87
Greenspan, Alan 50
Greg, W.R. 175

H

Haass, Richard N. 3
Halberstadt, Victor 3
Hamilton, Alexander 32
Harmonies of the World, The 186
Harwood, Richard 87
Hereditary Genius 175
Hitler, Adolf 7, 28, 54, 67, 111, 136, 181
Hoechst AG 66
Honeywell International, Inc. 199-202
Hoover Institution 105
HSBC 19, 28
Hubbard, Allan B. 3
Hudson Institute 105
Huxley, Aldous 9, 76, 162, 168, 173-175, 179, 182-183, 185, 192, 211, 220
Huxley, Julian 180, 182, 185
Huxley, Thomas H. 171, 174, 183

I

IBM 160, 169
I.G. Farben 66-67
IMF 48, 52, 153
Impact of Science on Society, The 89, 113
Imperial Chemical Inc. (ICI) 44
Inciarte, Matias Rodriguez 3
ING 19, 28
In-Q-Tel 102-103, 135-136
Institute for Creative Technologies (ICT) 147-148, 152
Inter-Alpha Group 19, 20, 29
International Herald Tribune 110
International Planned Parenthood Federation 182

J

Jeremiah, David E. 139
Jet Propulsion Laboratory (JPL) 146
Johnson, D. Gale 38
Johnson, Lyndon B. 26
Jones, James L. 3
JP Morgan Chase 60

K

Kaiser Wilhelm Institute for Genealogy and Demography 181
Kaku, Michio 155, 155
Kallmann, Franz J. 71
Kasner, Edward 99
Kellogg, John Harvey 180
Kepler, Johannes 186
Keynes, John Maynard 180
Kissinger, Henry 3, 43, 55, 164, 183
Kleinwort Trust 29
Kline, Nathan 188
Kohlberg, Kravis, and Roberts, (KKR) 103
KPMG 28
Kravis, Henry R. 3, 103
Kravis, Marie-Josée 3, 103
Kulcinski, Gerald 121
Kurzweil, Ray 164, 189, 220

L

Lamarck, Jean-Baptiste 180
Lanier, Jaron 190
Lazard Frères 111
Lehman Brothers 26, 60
Lewin, Kurt 7, 12
Licklider, J.C.R. 146
Limits to Growth, The 17
LinkedIn 87-88
Lloyds Bank Plc 28
Loeb, Jacques 70
Lombard Odier & Cie 43
Louie, Gilman 102-103
Louis Dreyfus 44-45

M

MacMillan, John Hugh 43
Maeterlinck, Maurice 92-93
Malthus, Thomas 175-177
Markle Foundation 102
Marrakech Agreement 38
Mass Psychology 79, 88
Matrix (movie) 205
Mengela, Joseph 181
Micklethwait, John 110
Microsoft Corporation 89, 94, 98
Milner, Alfred 171, 174
Minority Report 140, 200, 202
MIT Technology Review 134, 155
MK-ULTRA 188, 194-195, 220
Monsanto Company 43, 45-46, 54-57, 59
Monti, Mario 28
Morgan, J.P. Jr. 180
MySpace 87, 91, 97

N

NASA 123, 125, 145-146, 154-155, 160, 163
National Endowment for Democracy 105
National Security Agency (NSA) 102, 114, 213- 214, 222
NBIC (nanotechnologies, biology, information technology and cognitive technology) 159, 164, 170, 184, 185, 187, 190, 193, 206, 219, 220
Nestlé S.A. 42, 55
Netflix, Inc. 88

News Corp. 109
New York Times 110
Novartis Agribusiness 44, 46

O

Obama, Barack 3, 43, 98, 154, 195, 196, 220
On the Origin of Species by Means of Natural Selection, or the Preservation of Favored Races in the Struggle for Life 176
Ortes, Giammaria 175
Orwell, George 9, 12
Osborn, Frederick 73, 78

P

PayPal.com 96, 99
Pearson Plc 109
Petraeus, David 101-102
Philippe (King) 17
Philosophie Zoologique 180
Pilgrims Society 105
Pinterest.com 87-88
Pioneer Hi-Bred International Inc. 44
Planned Parenthood 181-182
Poindexter, John 140
Pol Pot 40
Project 1980s 18, 21, 22
Prometheus (movie) 165-166, 219
PROMIS 140-144
Pusztai, Arpad 56

R

RAND corporation 105
Rappoport, Jon 64
Raytheon Company 145, 160
Rees, John Rawlings 93-94, 183
Reuters 109, 113
Ringberg Castle 2-4
Roche Group 28
Rockefeller, David 3, 16-17, 21, 38, 50, 71, 164
Rockefeller Foundation 52, 54, 57-58, 70-74, 105, 110, 173
Rockefeller, John D. III 71
Rockefeller, Nelson 71
Rolls Royce Ltd. 28-29

Roosevelt, Franklin Delano 50
Rothschild, Jacob 19
Royal Bank of Scotland 19, 28, 30
Royal Dutch Shell (RDS) 28-29, 111
Russell, Bertrand 89, 113
Russia Today 95

S

Sandoz Inc. 44
Sargant, William 93
Schmitz, Hermann 66-67
Schwarzenegger, Arnold 150
Scott, Ridley 78, 165
Seeds of Destruction, The 47, 50, 77-78
Sirotta, Milton 99
Skype.com 99
Social Biology 181
Société Générale 19, 30
Sony Corporation 100, 107, 109
Spectrum Magazine 217, 222
Stanford Review 96
Strategic Trends 2007-2036 4, 7, 9-12, 40, 76-78, 103, 114, 122, 135, 139, 152, 155-156, 165
Sutherland, Peter D. 3
Syngenta AG 44

T

Tavistock Clinic 8, 84-86, 94
Teledyne Scientific & Imaging 199
Terminator 150
Thiel, Peter 96-97
Thomson Corporation 109
Thurn und Taxis 19
Time (magazine) 87, 107-108
Time Perspective and Morale 7, 12
Töpfer International 43-44
Total Information Awareness (TIA) 140
Tradax, Inc. 43
Trade-Related Aspects of Intellectual Property Rights" (TRIPS) 37
Trichet, Jean-Claude 3
Trilateral Commission (TC) 20-21, 38, 49, 164
Trist, Eric 6, 8-10, 83, 84
Truman, Harry 50
Tsarion, Michael 92

Twitter 87-88, 91-92, 94-95, 98, 113
Tyson Foods, Inc. 45-46

U

UBS 28
UC Berkley 185
UNESCO 180, 182-183
Unilever 28, 42, 55
United Africa Co. 42

V

Vance, Cyrus 21
Viacom 108
Vivendi Universal 108
Volcker, Paul 21, 22, 49

W

Wachovia Bank 19
Wallenberg, Jacob 3
Wall Street Journal 109, 208,-209, 221
Washington Post 87, 110
Wells, H.G 173, 183
WHO 64, 65, 66, 67, 68, 72
Wired 97, 101
Wolfe, Lonnie 84, 113-114
Wolfensohn, James D. 3
Wolfowitz, Paul 3
World Bank 3, 43, 48, 52-53, 153, 182
World Federation of Mental Health 181, 183
World Health Organization (WHO) 63-65, 72
World Trade Organization (WTO) 37, 38, 39, 41-42, 47, 53, 65, 77
World Wildlife Fund 181, 183

Y

Yahoo 105
YouTube 88, 99

Z

Zeneca Agrochemicals 44
Zoonomia 179
Zuckerberg, Mark 95, 96